T0136713

Bloomsbury Scientists

Bloomsbury Scientists

Science and Art in the Wake of Darwin

Michael Boulter

First published in 2017 by
UCL Press
University College London
Gower Street
London WC1E 6BT

Available to download free: www.ucl.ac.uk/ucl-press

A CIP catalogue record for this book is available from the British Library.

ISBN: 978-1-78735-006-9 (hbk)
ISBN: 978-1-78735-005-2 (pbk)
ISBN: 978-1-78735-004-5 (PDF)
ISBN: 978-1-78735-007-6 (epub)
ISBN: 978-1-78735-008-3 (mobi)
ISBN: 978-1-78735-009-0 (html)
DOI: https://doi.org/10.14324/111.9781787350045

In memory of W. G. Chaloner FRS, 1928–2016,
lecturer in palaeobotany at UCL, 1956–72

Acknowledgements

My old writing style was strongly controlled by the measured precision of my scientific discipline, evolutionary biology. It was a habit that I tried to break while working on this project, with its speculations and opinions, let alone dubious data. But my old practices of scientific rigour intentionally stopped personalities and feeling showing through. They cannot work here where there is so much colour in the uncertainty. So my old style had to go. Curiously, it was much harder to find facts from this 100-year-old history of local human culture than from the 50-million-year-old sediments of the European Tertiary. At first, that made me feel even less secure. So I would like to thank Rebecca Stott, professor of literature and creative writing at the University of East Anglia, for her help in trying to crack this old nut. I hope that it reveals a softer centre. I really enjoyed our teatime tutorials at the British Library.

Others who helped with earlier stages of the whole work were Biddy Arnott, Robert Harding, Gavin McCrea and Stephen Masty. Stephen Phillips, Geoffrey Vevers, Jane Simpson, Ann Steuve, Martin Pick, Raphe and Cathy Kaplinsky, Julia Tracey, Christopher Hourmouzios and Derek Winterbottom gave help with smaller parts of the story.

Jon Gill made me aware of the reincarnation of UCL Press, and Chris Penfold has been at the front of a wonderful team there. I thank them all for their hard work. Liz Hudson was a wonderfully vigilant copy-editor.

Very belatedly, I would also like to thank two of the long-deceased schoolmasters at my Leicester grammar school, Alderman Newton's: Bert Howard and the headmaster Hector Gaskell.

Contents

List of figures

Note: dates where known.

Chronology

1871 Lesley Stephen appointed editor of *The Cornhill*.
1878 Alfred Wallace joins Spiritualists in Upper Bedford Place.
1880 Thomas Huxley lectures on science education and social class.
1880 Ray Lankester, *Degeneration: A Chapter in Darwinism*.
1882 Lesley Stephen and The Tramps have tea with Charles Darwin.
1882 Charles Darwin's funeral.
1882 Matthew Arnold lectures on literature and science.
1883 Francis Galton Anthropometric Laboratory collects data.
1884 The Fabians are founded.
1884 Ray Lankester appoints Karl Pearson at UCL.
1885 Karl Pearson, Olive Schreiner and Bryan Donkin start the Men and Women's Club.
1885 Samuel Butler finishes *The Way of All Flesh*.
1887 Roger Fry elected to the Apostles.
1888 Roger Fry and H. G. Wells graduate in biology.
1889 Ray Lankester appoints Raphael Weldon at UCL.
1889 Julia Stephen signs *Appeal against Female Suffrage*.
1889 Arthur Tansley begins biology studies at UCL.
1890 William Booth, *Darkest England*.
1890 H. G. Wells publishes two magazine articles comparing art and science.
1890 Francis Oliver appointed professor of botany at UCL.
1891 Ray Lankester leaves UCL for Oxford University.
1892 Havelock Ellis, *The Nationalisation of Health*.
1893 Thomas Huxley gives Jack and the Beanstalk lecture.
1893 Ray Lankester begins Evolution Committee at the Savile Club.
1893 Benjamin Kidd meets Roger Fry and John McTaggart at Cambridge.
1893 H. G. Wells publishes first book, *Textbook of Biology*.
1894 Roger Fry appointed art critic for the *Burlington Magazine*.
1894 H. G. Wells, *The Time Machine*.
1894 Benjamin Kidd, *Social Evolution*.

1894   Ray Lankester engagement to marry is broken.

1895   Virginia Stephen becomes secretary of Stephen Entomological Society.

1895   Ray Lankester arrested in Piccadilly.

1898   Ray Lankester appointed director of British Museum (Natural History).

1899   Marie Stopes begins botany studies at UCL.

1900   Ray Lankester writes congratulations to H. G. Wells.

1900   William Bateson rediscovers Mendel's manuscripts.

1901   Arthur Tansley starts ecology journal, *New Phytologist*.

1901   William Bateson criticises Karl Pearson's manuscript.

1901   William Booth and Beatrice Webb collect data on London poverty.

1902   Marie Stopes graduates at UCL.

1903   George Bernard Shaw, *Man and Superman*.

1903   Brown Dog vivisection affair at UCL.

1904   Ray Lankester concerned at Eugenics Record Office in Gower Street.

1904   The four Stephen orphans move into 46 Gordon Square.

1906   Ottoline Morrell moves into Bedford Square.

1907   George Bernard Shaw describes Karl Pearson as a failure.

1907   Ray Lankester resigns from Natural History Museum.

1908   E. M. Forster at the Working Men's College, Great Ormond Street.

1908   Agnes Arber and Gwen Raverat suffer sexual discrimination.

1910   Walter Sickert painting workshop fearing sex, race, degeneration.

1910   Ottoline Morrell and Roger Fry explore modern art in Paris.

1910   Bertrand Russell and Alfred Whitehead publish *Principia Mathematica*, vol. 1.

1910   Maynard Keynes and Herbert Darwin begin Cambridge Eugenics Society.

1910   Ray Lankester begins Easy Chair column in *Daily Telegraph*.

1910   *Dreadnought* hoax.

1910   Roger Fry's Monet and the Post-Impressionists exhibition.

1910   T. H. Morgan, New York geneticist, finds genes on chromosomes.

1910   'On or about December 10th human character changed', Virginia Stephen.

1911   Virginia Stephen married Leonard Woolf.

1911   Rebel Arts Centre, 38 Great Ormond Street, Wyndham Lewis and Jacob Epstein.

1911   Henri Gaudier-Brzeska and Henri Bergson at Rebel Art Centre.

1912   International Congress of Eugenics, Hotel Cecil, Westminster.

1912   Ray Lankester frequent guest of Wells in Essex, Anna Pavlova in Hampstead.

1912   Second Roger Fry Post-Impressionist exhibition.

1912   Roger Fry and Ray Lankester contribute to H. G. Wells's *Socialism in the Great State*.

1913   UK Mental Deficiencies Act.

1913   Ronald Fisher's equations support gradual evolutionary change.

1913   Virginia Woolf's first novel, *The Voyage Out*.

1913   Adrian Stephen and Karin Costello at Rebel Art Centre.

1920   The Memoir Club founded.

1920   Marie Stopes, *Radiant Motherhood*.

1922   George Bernard Shaw's Lamarckian polemic *Back to Methuselah* first performed.

1926   Ray Lankester attacks Conan Doyle's *History of Spiritualism*.

1928   Gip Wells appointed an assistant in zoology, UCL.

1931   Julian Huxley, Gip Wells and H. G. Wells publish *Science of Life*.

1931   Tots and Quots begins meeting, Great Portland Street.

1932   Aldous Huxley, *Brave New World*.

1932   Julian Huxley appointed secretary, Zoological Society.

1937   Theo Dobhansky, *Genetics and the Origin of Species*.

1942   Ernst Mayr, *Systematics and the Origin of Species*.

1942   Julian Huxley, *Evolution: The Modern Synthesis*.

Introduction

This is the story of the network of artists and scientists living in a square mile of London just before and after the First World War. The group included men and women who made hard discoveries in biology while others created different kinds of art. It was a time of unexpected opportunities from new disciplines such as genetics and ecology, from the new art of post-impressionism and from a style of writing called the stream of consciousness. New technology was transforming family life just when more women were reacting to Victorian rules and influencing a new society. The physical sciences had already done well to improve standards of living and many asked whether it was the turn for biology to make another kind of impact.

Big social changes were also under way as power was being transferred from the landed oligarchy to some of the high achievers who had both ability and an education in science. New movements supporting socialism and nationalism were growing in response to the spreading urban poverty and a declining empire, but the controls in breeding that were being suggested by some members of the Eugenics Education Society were not going to be popular with many of these artists and scientists who were then in Bloomsbury. Nevertheless, the society had some surprising supporters.

The scientists and artists of this story brought their creative energy together to drive this modernity, working in the laboratories, libraries and studios of Bloomsbury, and they were more than just the sum of these parts. They were a unit observing the changing social mood of Edwardian Europe, describing its forms and feelings and recording these with a careful choice of words and images. This all gave a beauty and a new synthesis of knowledge to ideas in science and art, and it led to different kinds of human relationships.

I was born in Leicester in 1942, into a working-class family and a life made hard by war and the lack of opportunities. For more than a generation the people I grew up with had been in the wrong place at the wrong time and their lives didn't show much improvement through my childhood. They were all wounded, physically or socially, and the rationing went on long after the war had ended. Whole communities missed out, with no chance of education and with modest expectations held down by a loyalty to the nation-state, tragic victims of both world wars. The city's streets seemed empty all the time, the buses ran exactly on schedule and we all heard the factory hooter at eight and then at last at six. For more than a decade after the war there were no signs of change, and I was expected to get a job in the hosiery factory down the street, with my mates.

My father, who had been wounded at the battle of the Somme, looked after the spare parts for Spitfires during the Second World War and then, when it had ended, scratched out a living by selling the scrap from local factories. He used to bring home the sacks of unused strips of canvas from the inside of tyres, then sort them into different shapes and sizes and sell them on, mostly to the shoe factories in Northampton. It was a treat to go and help in the canal-side barn he rented, staying warm by a little stove in winter and being cooled by a breeze through the wide doorways in the summer. To fold the canvas into bales we would stretch the fabric sheets along the length of the towpath beside the barn. We had to struggle to keep them away from the water when the wind was in the wrong direction.

On Monday evenings, for a time, my mother took me to Mrs Crutchley's little terraced house in Walnut Street for piano lessons so that she could work at a nearby factory for a few extra shillings. I liked to play the piano. I used to dream that I would one day play at the concert parties that were popular in the 1950s. (We didn't know then that this trend would soon be replaced by rock and roll.) But I didn't like Mrs Crutchley. Whenever I made a mistake she would flick a long pencil across my knuckles. I wanted to escape from that cold room – and from the prospect of working in a factory in Leicester for the rest of my life – but such an escape would require a higher education, and my family could not afford to send me to a private school. I was going to have to win a place at university.

A lorry came for my father every morning, dead on time, and they filled the back with sacks of scraps and waste that some new production line had left behind. He could salvage stuff from army surplus outlets and sell it on if he was lucky. I preferred my books. We had only four at home: *Treasure Island*, the *Observer's Book on Pond Life*, an encyclopaedia and one about gardening. Through these books I came to sense the larger world beyond Leicester. We knew so little about what was happening outside our limited scrapyard and factory-oriented routines.

One day my father had the idea of using his strips of canvas to make fishing-rod bags. He cut out the cloth, my mother stitched it on her old Singer sewing machine, and they sold them directly to anglers through the classified advertisements in the *Angling Times*. This cottage industry did quite well, and our trips to the canal-side barn became more frequent. More exciting still were other journeys to fishing tackle shops in cities nearby where we sold the homemade bags by the dozen. But within a year or two my father's health worsened, and the division of labour had to change. It became clear that I was going to have to contribute to the family income. It had always been assumed that I would go into the hosiery factory, but now my parents began to wonder whether I should become a salesman in the fishing tackle world. Already, in my teenage years, I was doing well at that work.

Films such as *Saturday Night and Sunday Morning*, *A Taste of Honey* and *A Kind of Loving* offer accurate pictures of childhoods like mine. The strong community spirit of northern towns in England gave support when it was needed. The films were also authentic in their portrayal of bright young people escaping down to London, and thousands of us did that. Many came with strong social and political beliefs and considerable financial help from the state that paid college fees and rent. My belief was that through science I would find a meaning to life: a means of escape from the past and a purpose to my own existence.

My parents had missed out on chances in life, and they were determined I should not miss mine. That meant I should have a better chance through a good education. The money from the salvaged scrap and fishing-rod bags was recycled into more books for my shelf. Even the local vicar helped, unknowingly, with the challenge that he would attend my ordination into the church if I kept working hard. Through those books I became more interested in natural history than hosiery or fishing tackle, and my grammar school helped me to focus my mind on the study of science. My education was conducted in the spirit of the Nonconformist tradition of social progress. I even had the same history master, Bert Howard, who had helped the novelist and science administrator C. P. Snow get a place at university more than thirty years before. Howard was a quick-witted Mr Chips-like intellectual who rated the underdog more highly than the establishment.

In his 1956 novel *Homecomings*, one of Snow's major characters, George Passant, is based on Bert. George, a solicitor's clerk with no prospects in our familiar provincial town, is interviewed by a senior civil servant for a job in Whitehall:

> 'Forgive me interrupting, Mr Passant, but with a school record like yours I'm puzzled why you didn't try for a university scholarship?'
> 'If I'd known what they were like I might have got one', said George robustly.
> 'Leaving most of us at the post', said Rose with a polite bow.
> 'I think I should have got one', said George, and then suddenly one of his fits of abject diffidence took him over, the diffidence of class.
> 'But *of course* I had no-one to advise me, starting where I did.'[1]

From my teenage years onwards, my interest and aim was clear: to leave Leicester for London and to become involved in the exciting new work in science and art that was taking place there. For my generation these were times of hope, and, unlike George Passant, who didn't take the chance, I was determined to move on. I won a place at UCL to read biology. My parents, unaware of its significance but full of pride, booked me an open return train ticket to London.

The train from Leicester pulled into St Pancras station, where that grand building was a more confident place than any I'd experienced before. Its massive smoky atmosphere went on into the distance, leading to the Gothic spires of the cathedral-like façade. Here, it seemed that everything could be challenged, and I could see in that one building so many new ideas. I was learning that a critical approach to established orthodoxies was essential for progress. With my single suitcase from a previous life, my undergraduate years would force me to relinquish many of my certainties and assumptions. More than anything the discoveries made in the first half of the twentieth century in Bloomsbury forced me to see living systems in a different way.

On my arrival in London, I was already familiar with the continuing series of New Biology paperbacks published by Penguin Books and largely written by Bloomsbury biologists from the early twentieth century to the 1960s. These books summed up what biology was about at that time: scores of different topics by young researchers from different backgrounds. Taken as a whole, they were a testament to the excitement that my new student friends and I were feeling then; we were witnesses to the fact that something special was happening. We all felt that by some strange coincidence we had landed in a very special time when things were changing for the good. The New Biology series showed that from the new fields of molecular biology and continental drift, Darwin's theory of evolution by natural selection was being proved.

During my first term in London I went to the theatre fourteen times – I have kept the programmes! I love the theatre, and today I continue to be moved by the feelings actors can transmit from the stage. *Beyond the Fringe* was much more in tune with where I was going than that glitzy pop culture of *My Fair Lady*, but wherever the productions took us there was always controversy within my group of friends about a production's emphasis and whether the atmosphere and words were appropriate to the main theme.

At UCL, the scientific dramas that we heard or overheard in the corridors were easier to interpret because, we thought, there was a right and a wrong way of interpreting the plot. We were also a privileged audience. Our lectures featured performances by the stars of molecular biology, Francis Crick, Jacques Monod and J. B. S. Haldane. Not only did I share the canteens and laboratories of Gower Street with these great scientists, but I was there at a time when Darwin's ideas were up for their final and most crucial test: would DNA have the right codes for Darwin's kind of evolution? We'd all been taught at school to suspect they would, but we weren't sure, and these men sensed that the answer was at their fingertips.

The stage could hardly have been better placed: a newly built lecture theatre on the very site of Macaw Cottage, the house in Gower Street where Charles Darwin and his wife Emma first lived together in the 1840s. It was in this house, at that earlier time, that the opportunities of adaptation were forming in Darwin's mind and where he sketched his first outlines for his theory of natural selection. Macaw Cottage was close to where many intellectuals would live during the next century. Down the road, in Bedford Square, Ottoline Morrell held her parties before the First World War and, just behind, on Gordon Square, Maynard Keynes and various Stracheys moved in during the 1900s. To the front of the house was Fitzroy Square, and there are now blue plaques to mark the homes of George Bernard Shaw and Virginia Woolf.

This square mile of Bloomsbury extends from the British Library and three great railway terminals on the north side to the portico and atrium of the British Museum to the south. In between, on Gower Street, is UCL, just beside Birkbeck College and the Art Deco Senate House of the sprawling University of London to the west of Russell Square. About ten other squares with plane trees and lawned gardens are scattered around that same area, and with lots of other students I lived for two years over-looking Cartwright Gardens.

Here I was in the 1960s, at the centre of these things. I was away from the salvaged bags and hosiery and into the bright new generation of Bloomsbury scientists. I was taking over from where these well-known intellectuals had left off before the Second World War. Sure enough, some of my new friends were from those intellectual classes, from well-off families in southern England with trendy clothes and expensive habits, there to study things like law and economics. I was even invited to weekend house parties in Oxfordshire only a few miles from where the Bloomsbury group had their own retreat at Garsington.

However, I was more at home with my scientific colleagues, who, comfortingly, were mostly from backgrounds like mine. Many came from the north, and they studied for long hours, with only thirty-minute breaks at teatimes and a day off on Sundays. Some of these new friends had just returned from the final year of National Service in the armed forces. We felt they were real men, classless, with broad interests, and they talked with experience and knowledge of all those things they had done and the foreign places they'd visited. The member of our group with the most money was Pete Bennett, a medical student funded by the army, who bought the biggest rounds of drinks in the pub and often went off to Germany on training exercises. Another was his schoolmate from Manchester Grammar whose harsh upbringing had taught him to rebel against the church and become a kind of born-again atheist. We were led by Ivan Vaughan from Liverpool, an economist in a black corduroy jacket, who personally knew The Beatles; indeed, he brought them down to stay one weekend. We all went to the pub in Lamb's Conduit Street and they slept on Ivan's floor the night before they went to play at the Star Club in Hamburg.

Each term there seemed to be discussions and parties everywhere; people mixing, arguing and dancing. Artists at the Slade smashed their guitars on the floor, and medics in Mecklenburg Square talked about Bach. There were arguments, and meetings of student societies, and newspapers and posters and circulars. All were excited by science and art and anything different to the way things were at school or before the war – or in Leicester.

This was how I imagined Virginia Woolf fifty years before, then known by her maiden name, Stephen. With her sister, two brothers and their many friends, she had lived at 46 Gordon Square. These four well-off children and their friends had escaped a very different kind of incarceration to mine, that of the Victorian establishment. In 1963, their house was being turned into the University Computer Centre and the bombsites were being redeveloped with new buildings for the rapidly growing university. The proud utilitarianism of Jeremy Bentham was protected in all the ceremonies of University College, which he helped to found. He was so much part of the fabric that it was assumed his consideration for all people would continue in this place, but for how much longer could his Nonconformism cultivate eccentricity and dissent?

In London we began to feel part of a machine for recovery, with our whole generation of young people slowly moving out of war and hardship into growth and a better life. It was what our parents had been fighting for, and we were confident and driven by hope. We felt real excitement hearing about the scientific breakthroughs as they happened. Even things that turned out to be failures thrilled us. The major discovery of these times was how DNA was coded to make proteins, and we knew, theoretically, that this could lead to direct evidence that Darwin was right. However, to demonstrate that idea experimentally was another matter. Famously, Watson and Crick had elucidated the double helix structure of DNA in 1953, and they went on to gather a team of biologists to work out how the genes on those long molecules programme the synthesis of enzymes. These are the proteins that drive biochemistry in every cell of every organism: in short, they run biology, and to understand how they come about is key to knowing how life works.

A famous molecular biologist, Sidney Brenner, lectured us one week about the new evidence he had found that linked DNA with protein synthesis. The next week he arrived to announce that his theory had been proved wrong a few days before because he had missed out a whole sequence of reactions. It was the best lesson in how science works that I have ever had. On another occasion, the geneticist David Wilkie began a lecture with the surprise news that he had just found DNA outside a cell's nucleus, in the mitochondria. This discovery broke all the established rules about how these molecules of inheritance work. We believed that the mitochondria were all about housing the biochemistry of respiration and had nothing to do with genetics. We were wrong. Wilkie's science won the day.

In the 1960s, geologists proved the existence of the continental drift of the earth's tectonic plates and refined accurate methods of dating the age of rocks. For thirty years, since Alfred Wegener's supporters had chronicled distinctive and unique fossil plants from the southern hemisphere, there had been talk of a southern land mass called Gondwanaland that some claimed had broken up 100 million years ago. Some of the geology lecturers had access to data from the sea floor of the Atlantic suggesting that sediments along the mid-Atlantic ridge were relatively young, only a few million years old, and their age got older the further away from that ridge, both to the east and to the west. The simple explanation was that North America was moving away from Europe, a few centimetres a year. Attention shifted to Iceland, sitting astride that ridge and getting bigger. As this group of scientists gained success and fame, a new generation of off-shore explorers began to make new discoveries about the history of the earth and the life that had evolved on it over millions of years, and so began the growing interest in changes of the environment and its climate.

Through all these years I kept up with my family duties in Leicester, helping out with our thriving cottage industry and keeping in touch with the friends I'd left behind. I often called in to see Bert who, every Wednesday evening, held forth in the men-only bar of the Saracen's Head, just off the marketplace. A group of six or more of his students was usually present. Often the Oxford historian Jack Plumb and Harry Hoff (the novelist William Couper) were there too, while on one occasion I met C. P. Snow himself. There was a lot of talk about time spent together on the Norfolk Broads where we were members of a sailing club, and once there was an argument about history and science and whether they worked in the same way.

My own budding interest in palaeontology meant that I remembered one evening, in particular, when Jack Plumb talked of the recent death of his supervisor, the great Oxford historian G. M. Trevelyan. Both Plumb and Trevelyan had seen history merging with some of the newly popular sciences of the times. In their own work, they had linked history to anthropology, archaeology and palaeontology, as well as to psychology and sociology. This was just after F. R. Leavis's 1962 Richmond Lecture at Cambridge in which he had attacked Snow's earlier critique of education in England.

All this movement away from home showed me that a deep intellectual divide was separating my family and my new Bloomsbury friends. Aware of this split, my ever-loving parents were thrilled at the thought of my escaping 'to do science in London', even though it meant my leaving. In turn, I was thrilled to avoid what I saw as the trap that had caught George Passant and to find some way of staying on in the world of science.

When I graduated, I knew that my knowledge from those early days of molecular biology was a marketable commodity, so I made a list of the local colleges that taught science. With a pocket full of old pennies I went into a phone box in the Russell Square underground station and called around, struggling to hear through the noisy rattle of the gates to the station lifts. An hour later, I had found myself four one-day-a-week jobs teaching DNA and the genetic code in London technical colleges.

I was to follow in the steps of many young scientists from Bloomsbury, teaching and researching evolutionary biology. The knowledge and methods used by the earlier Bloomsbury scientists were greater and very different to mine, and their society had different problems and concerns, but there were also many similarities. Now, another fifty years on, my work from those years has as much historical value as scientific. Altogether, the barren years for biological research from Darwin's death to the 1950s, and the great breakthroughs of the 1950s and 1960s, marked a most exciting time for the life-scientists privileged to be part of those explorations.

1
Two funerals, 1882–3

Ray Lankester, a young professor of biology in Bloomsbury, went to two landmark funerals within a few months. Both were of great public intellectuals and would have reminded him, if he needed reminding, of the immense diversity of intellectual life in London and of the intricate and overlapping networks in which he lived.

The first funeral, at Westminster Abbey in the spring of 1882, was Charles Darwin's. The congregation lined up along the nave of the Abbey, an impressive display of the great and the good and the well-connected, at a time when the British Empire was at its most powerful. Lankester was thirty-five years old and had been a professor of zoology for eight years. He realised more than the others how important Darwin's academic and social legacy was. He would have looked up and down the rows of scientists in the congregation for a likely successor to the great man, and he would have known how difficult it was going to be for anyone to take on that mantle. Many of the upcoming zoologists present may have been considering their own positions in the scientific hierarchy, wondering if they would ever receive such adulation.

Lankester had been invited to sit in the nave next to Sir Leslie Stephen and other gentlemen scientists but they found conversation difficult. Like Darwin, they were pioneers in many fields of science, all fellows of the Royal Society and devoted to the exciting challenges of the new biology. As well as Stephen, there was Darwin's cousin, Francis Galton, from a Quaker family of bankers and gunsmiths and able to fund his own interests in many diverse scientific projects. Then there was Richard Strachey, just back from a career in the Indian civil service, building railways and still describing plants he had collected from the Himalayas.

These men enjoyed the status and power of their roles in science and saw no reason why they should relinquish their positions for any challenge of accountability. It was accepted that the young Ray Lankester should be the last in their line walking out from the Abbey.

However, Lankester was different to these Victorian gentlemen. He was one of the small number of publicly appointed science professors and came from a different background to these middle-aged landowners, civil servants and investors. Maybe that was why he was particularly good at spotting creative talent in his students and fertile topics for research in the new biology, applications to improve public health and reduce poverty. That was how he had become one of the few leading scientists well aware that biology was a big political issue, offering an alternative to God as well as explanations of the different races and of how to improve the quality of human health.

Karl Marx's funeral, the following winter, was a small and bleak affair attended only by Lankester and ten other mourners. The style and atmosphere of that ceremony was in stark contrast to the one at Westminster. It took place on a chilly hillside at Highgate Cemetery where Lankester stood with Marx's daughters, Laura and Eleanor, and the factory owner Friedrich Engels. The others were a motley crew of political supporters just released from prison in Germany and France, and Karl Schorlemmer, a professor in chemistry from Manchester. Lankester had taken a horse-drawn tramcar to Kentish Town and walked from there to the cemetery. After a short ceremony, the coffin was taken off its four-wheeled carriage, and the small group, long black coats billowing in the open air, watched it being lowered into the ground.

Lankester had known and admired Karl Marx for the previous three years and had recently helped the family through some hard times. Lankester and Marx had talked together about their own work and had concluded that their apparently different academic interests had much in common, especially when considering the question how individuals behaved in groups and how those groups responded to change. Darwin's theories had affected them both, and, though they disagreed about how change is brought about, they had high regard for each other's intellect.

As the mourners walked away down the long drive, Lankester had a chance to talk to Engels for the first time. During that conversation, Engels very likely explained that Marx had been influenced by the discussions he had with Lankester about evolution and science. It also surprised Lankester that Engels, the industrialist, had such a clear understanding and grasp of developments in biological theory, and how much of this thinking Marx had incorporated into his own political theories. Lankester had always emphasised to Marx how important it was to understand that both nature and human life are in a continuing state of change, so he was interested to discover that Marx had once presented in a letter to Engels an impressive interpretation of Newton's influence. 'Motion is the mode of existence of matter', Marx had written. 'Never anywhere has there been matter without motion, nor can there be.'[1] In another letter, Marx had argued that living systems depend on the interaction of cause and effect, and Engels thought this 'went more to the heart of the matter than any of the empirically minded English scientists could possibly appreciate'.[2] Evidently, political economics and evolutionary biology had more in common than anyone had previously thought.

Lankester went on to talk to Engels of his recent thoughts that ethics were not a feature of evolutionary change. Marx had in mind a different kind of struggle for existence, but Engels and Lankester did not like to criticise the dead man at his funeral. These were the same issues that preoccupied Lankester and his friends in Bloomsbury and Kensington. So far, Lankester had done more than most to talk publicly about the implications of Darwin's work. The two funerals had shown him just how broad the importance of Darwin's work had become. There was a lot of unfinished research to be done and new evidence to be obtained before the full scope of adaptation and natural selection could be known and accepted. Lankester knew he was a good anatomist and microscopist, and his role as editor of the *Quarterly Journal of Microscopical Science* meant he was in a very good position to understand evolution. But much more was needed.

At the funerals of Darwin and Marx, there were two different groups of thinkers: one was rich and Oxbridge-educated; the other was dispossessed of public responsibilities, rejected authority and wanted to change society at all costs. Despite being in opposing political camps, both groups believed deeply that science would solve the world's problems. They represented the two extremes of the social turbulence that was already gripping Britain and that was to develop and consume Europe thirty years later. Between these poles were young people from different backgrounds, education, religious faith, artistic taste and scientific interest, and they were changing their social attitudes very quickly.

This young generation had grown up to respect gentlemen scientists such as Stephen and Strachey, but the lifestyle was rarely sustainable, and open competition for jobs meant there were growing numbers of applicants for many positions of responsibility. With his developing spirit of hubris, Lankester was one of these, now aware that he was at the leading edge of many of these changes, both as a scientist and as a university teacher. He was finding meaning to life processes and explaining them to leaders of the next generation. What better place to do this than at UCL, in Bloomsbury, founded on Jeremy Bentham's utilitarian belief in the greatest good for the greatest number. Bentham had called Oxford and Cambridge the 'two greatest public nuisances, storehouses and nurseries of political corruption'.[3] Membership of the Church of England was still necessary for admission, so all Nonconformists, Catholics and Jews were not allowed that education.

Darwin's *On the Origin of Species* offered an alternative to religion to explain life's origin and meaning. Technological developments were making it much easier to undertake physical tasks, and it was becoming more acceptable to rebel against the rigid hypocrisy of Victorian society. However, this was England, and there was also the simple matter of social class. Many years later, Stephen's daughter, then known as Virginia Woolf, described its force between two of the characters in her novel, *The Years*:

> 'He was not Eton or Harrow, or Rugby or Winchester, or reading or rowing. He reminded her of Alf the farm hand … who had kissed her under the shadow of the haystack when she was fifteen.'
> To which came the reply: 'She's a stunner, he said to himself, but my word, she gives herself airs.'[4]

***

Since the 1870s, the ambitious Ray Lankester had rented rooms on the ground floor of a house in Kensington, by Hyde Park. This was where many of the scientific establishment had their London homes, and he shared a housekeeper with one of the other tenants. This was a common arrangement for 'respectable' single men, one that saved them from housework and cooking. He was around the corner from the Stephens's at Hyde Park Gate and Francis Galton at Rutland Gate, while the Stracheys were just across the park. It was easy for him to cultivate the image of a gentleman with his warm personality and impeccable manners. Like many bright young men of his new generation, Lankester travelled regularly to places such as Oxford, Plymouth, Italy, France and Germany, but such a lifestyle, and the discrepancy between his cultivated image and his social reality, made him very lonely. It was hard for him to make friends with people outside his academic circle.

Figure 1.1   Edwin Ray Lankester

Throughout the 1880s, the Stephen and the Galton families had the servants and the houses that befitted their senior status. The rest of the household, their children and the servants, were all squashed up in other parts of the six-storey buildings. There was no electricity, and the coal had to be carried to every fireplace and then cleaned out the next morning. The one flush lavatory made a loud clanking noise, and the cast-iron cistern was often out of order through overuse. Home was not the luxury that many observers have since imagined; it was, rather, a cold and noisy space where it was hard to relax or be intimate.

Lankester was thankful to be out of all this and able to spread out his books and articles on the table in the front room where only he knew they belonged. This seemed essential for him to think, and it was also common for him to read at least one book a week or to delve into one of the many weekly magazines that serialised the new fiction. The best of these was *The Cornhill Magazine*, which Stephen edited.

Indeed, when Lankester thought of culture he always thought of Leslie Stephen. The man exemplified the social and cultural circle around which so much of London's artistic and scientific life then turned. Although he wrote about the ethics of scientific practice, Leslie Stephen did not have much to do with scientists, and he described the few he had ever met properly as 'utterly inscrutable fanatics'.[5] In turn, Lankester was suspicious of Cambridge clerics, especially those who had renounced their vows and turned from teaching to journalism to make a living (although Stephen's recent work on ethics just about redeemed him). While following a scientific career, Lankester wanted his work directly to improve the living conditions of the poor, and he believed the key to this was the use of science as an ethical guide. It had become his new religion and was forcing him to reshape the religious morality with which he had grown up.

Like Darwin, Leslie Stephen was an amateur scholar and a Victorian gentleman. He had been brought up in Clapham where the evangelical church was especially strong, but he was impressed by Charles Darwin's one grand evolving system for all living things. Back in 1854, when he graduated at Cambridge, there was much less evidence about evolution. The argument for adaptation by natural selection was strong, even though the evidence was circumstantial. All that Stephen knew about his own future was that he did not want his easy lifestyle to end, so he applied for and secured a fellowship at Trinity Hall. The duties of this post involved giving tutorials to the new students at the college and taking holy orders to continue its religious traditions. It was a clever move because it gave him an easy life, even though he wasn't sure that his faith was strong enough. After several weeks of uncertainty, he was eventually persuaded to take holy orders on the grounds that he would remain faithful to the traditions of rural England, traditions founded on a deep love of nature and its beauty. More importantly, it also meant that he could carry on rowing and enjoying nature and staying in the good company of college men.

After almost a decade of this so-called 'tutoring' at Cambridge and climbing mountains around Zurich, Leslie Stephen became tired of whiling away his time in carefree hedonism. He married, found a job as a journalist and settled down in a large house in Hyde Park Gate. By 1871, he had become editor of *The Cornhill,* the leading thinking-person's magazine, to which he was able to attract new writers such as R. L. Stevenson and Thomas Hardy. His regard for most of these people usually remained strictly professional, though occasionally they became close personally. Stephen and Hardy met frequently over lunch, snooker or dinner at their club, the Savile. They were part of a group there that included Lankester and R. L. Stevenson, and they all supported Stephen when he began to question his own religious beliefs.

With time, Stephen made a series of unexpected and important connections with other scientists and artists. Some of them had helped with his major work over those years: *The History of English Thought in the Eighteenth Century*, published in 1876. This gave a slow and peaceful reaction to the Enlightenment and a lot of space to the Industrial Revolution. He was forming a complex web of families, specialists and friends, who held new ideas about liberty, human rights and the relationship between science and society. He became more concerned with human behaviour and morals than with art for art's own sake, and he wanted a new set of rules at work and at home: a new structure to his life. However, his was still a religious world, and, just as he had taught theology with the authority of a preacher who expected his students to believe and to obey, now he expected his readers to *learn and inwardly digest*. He could not quite let go of his evangelical roots, so he thought long and hard about his allegiances. This was what most of his friends called atheism with a human cause, a particular kind of ethical guideline.

Stephen's religious beliefs were cruelly tested in 1875 when Minnie, his first wife, died in childbirth at the age of thirty-eight. This event left Stephen depressed, angry and lonely. A few years later he married Julia Duckworth, a neighbour who was also widowed and with children. Julia challenged Leslie with her anti-Christian sentiments, and she gained confidence from the social round of like-minded friends in London society. Many of those also challenged religion and thought that science was going to be the driving force behind the most important challenges in the foreseeable future.

Their own ethical guidelines were the foundations of Victorian England, at a time when many other leading thinkers were drawing attention to the different political policies as alternatives to traditional ones. As we shall see, other new thinkers replaced God with Good and Christianity with Morality. Evolution became the Creator, with man as his child. As the editor of *The Cornhill*, Stephen was influenced by many of his authors, and some of them became his heroes. However, through it all, with his competitive Etonian spirit, he tried to go one better himself with examples of high morals, modernity and scientific ways of working.

All these thinkers were excited in the belief that their science of evolutionary biology was offering an alternative set of codes to replace some of the teachings of the church. Some agreed with the sociologist Herbert Spencer who saw evolution leading only to more complexity and who argued that, if this happened in new species, it would also apply to new societies, leading naturally to more civilised behaviour with a more complex moral code. Most accepted Darwin's ideas, though what he wrote about race and sex were topics that most Europeans found difficult to discuss.

Another of Darwin's influential books, *The Descent of Man*, had been published in 1871 and argued how different human races could be regional variants of our single living species. We were all adaptations to different environments and all from a population of common ancestors living several millions of years ago. Darwin also had a clear but impractical solution to the problem of an increasing birth rate: 'Both sexes ought to refrain from marriage if in any marked degree inferior in body or mind.' He saw no other way to avoid 'abject poverty for themselves and their children'.[6]

So began the ideological debate about how to respond to differences in the variations and adaptations in racial features of the human species as well as in our vast range of physical and mental attributes. The arguments touched on other issues such as the sustainability of human populations – questions that we continue to worry about today with no sign of resolution. Through the 1880s, Lankester responded by lecturing and writing about the dangers of degeneracy and of rewarding the rich regardless of talent. He stressed the importance of loving care within families and how abandoning hereditary privilege would help strengthen the natural population. With the same respect for nature, he feared that state regulation about reproduction would lead to artificial pressures within the natural population. The more regulations imposed by politicians, the more difficulties would be created.

By the spring of 1875, Stephen decided to renounce the vows he had made at Cambridge. Significantly, it was while he was writing *History of English Thought in the Eighteenth Century*. Without giving any hint of his purpose, he summoned his close friend, Thomas Hardy, to the house in Hyde Park Gate. He asked Hardy to arrive late at night when most members of the household would be asleep. Hardy saw his host as 'a tall thin figure wrapped in a heath-coloured dressing-gown … wandering up and down in his library slippers'.[7] After the two men had taken a glass of brandy, they went up to Stephen's library where he had laid out a parchment document on the candlelit table in the middle of the room. The document was a revocation of the holy orders that Stephen had vowed at Cambridge twenty years before. It was to take Hardy's signature as witness to this important change of heart. The two men shook hands in approval.

From that moment, Stephen took his agnosticism seriously. Late in 1876, he was seen standing on a soapbox beside the front entrance to Lambeth Palace, the official residence of the Archbishop of Canterbury, telling the multitude why he left the church. The scene 'was exactly like a Holbein – the magnificent head, with its strong red hair and beard, painted against the porcelain-blue sky ... He seemed like a prophet raised half-way to heaven, high above the people.'[8] With that confidence, and encouraged by Darwin's *The Descent of Man*, Stephen won the support of many of his scientist friends such as Thomas Huxley. At the Savile, he conversed with Lankester and Stevenson, and, instead of going to church, he joined a group of nature lovers called 'The Tramps' who went on a country walk every Sunday. This was the group that Stephen arranged to take to Darwin's house in Kent for tea in January 1882, just a few days after the birth of his second daughter, Virginia, and a few weeks before Darwin died. The former priest recalled that Darwin's *On the Origin of Species* had just been published when he had taken his vows in 1859. Stephen said that, since he did not have much knowledge of biology, he was impressed more by the book's enthusiastic reception by biologists than by its argument. As his scientific knowledge deepened over the next few years, he became ever more convinced that organisms are part of the same evolutionary tree, adapted to environmental changes. These thoughts were strengthened by the conversations with his second wife, Julia, and he used them later to encourage their children to be interested in natural history. It was so different from the years of his life with Minnie, when he hid away in melancholy and intellectual loneliness. However, it turned out that only his own death would release his four children from that dark place in which they were brought up.

Julia's ideas about womanhood also had a significant impact on Stephen's understanding of biology and its relationship with human society. She believed that the maternal influence of women improved the kindness between people, especially relatives, and it also diminished male egoism. Although a supporter of science and promoter of technology such as electric power, she could not reconcile female emancipation with motherhood and home life. Along with most of her friends, Julia believed that her role was a domestic one. A woman's task, for her, was to stay in the home and give patient support and attention to men.

However, there was an antithesis growing in London society, and Julia was one of many women of her generation who supported it. In 1889, Julia signed the *Appeal against Female Suffrage* that had been put together by her friend Octavia Hill. It was soon to be taken up by none other than the founding Fabian, Beatrice Webb. Hill was also a busy social reformer in London throughout the 1870s and 1880s, moving people out of slums and saving green spaces, such as Hampstead Heath, from the developers. Lankester would have known these women through their work to reform post-school education, in particular their involvement in the Working Men's College in Great Ormond Street. So keen were they to help educate working women that they set up a neighbouring Working Women's College with many of the same staff. Hill became a regular teacher there and helped Elizabeth Garrett teach zoology with live frogs. Amelia and George Tansley, who lived close by in Regent Square, were also enthusiastic students and became occasional teachers at both colleges. It was an early sign that some different kinds of people were joining together to improve their lifestyles and those of their neighbours.

By 1885, Leslie Stephen was not entirely satisfied with his new way of life. He was editing *The Dictionary of National Biography* and writing many of the entries himself. He was working long hours at the British Museum reading room and had tight deadlines and few breaks: 'I am knee deep in that damned Dictionary and drudgery.'[9] Julia hated the *Dictionary* and its demands, suspecting that his sleepless nights were leading to a breakdown.

Many of those writers who knew Stephen wrote about his move away from religion. There are several references to biological detail in the literature of the period, the very books that were being read aloud in the drawing rooms of Kensington. For instance, George Eliot, partner to the polymath physiologist and philosopher George Henry Lewes and an enthusiastic follower of Thomas Huxley, had long been structuring the great philosophical questions of her novels in terms of the debates current in biology. Doctor Lydgate in *Middlemarch* espoused passionate commitment to advancements in biology as a way of ensuring medical progress. Aware of the debate about biological evolution and creationism, Eliot pointed out in the preface to *Middlemarch* that 'the limits of variation are really much wider than anyone could imagine'. She went on to take that idea further: 'Art is the nearest thing to life', inspiration that caused Virginia Woolf to say later that *Middlemarch* was 'one of the few English novels written for grown-up people.'[10] It was also more early evidence that a new culture was developing with the influence of science.

Here was a new ethical question: How do humans fit into nature? We were the only species able to seriously change the environment, thereby having direct influence on the lives of other species as well as our own. Stephen particularly liked the way Eliot, in her novels, showed natural diversity within human societies. This was something that Thomas Huxley also took very seriously and wrote about at length. One of his popular metaphors showed nature as a garden that needed regular human attention in order to keep it in good shape. The denial of the normal processes of nature was an essential part of what Huxley, in particular, regarded as human: the human propensity for designing and making gardens as opposed to allowing nature to take its own course. He was sure that ethics and morality were also things to be devised by humans, not by nature. For Huxley, ethics and morality were to be created and revised as society saw fit. However, there was a conflict between being good for the sake of self-improvement and the opportunist struggle for existence. To better oneself, one might have to be nasty.

These late-nineteenth-century debates about biology were bound up with questions about sexual politics, ethics and the roles of men and women. The debates took place especially in those houses where there was an attempt to live free of orthodoxies and to understand life in terms of biology rather than religion. These new attitudes of the Stephens were shared by many of their friends, including Jane and Richard Strachey who lived across the park in Lancaster Gate.

Jane Strachey believed deeply that science could help involve women in a better lifestyle. Despite the difficulties of working in India, having a house in London, being mother of nine children and wife of Richard, Jane was not one to stay at home and be quiet, and she developed her own serious interests in science and scientists. Richard became well known as a man interested in 'real problems'. He was an engineer, building railways in India, and his surveying techniques allowed him to make observations of biology and the environment, which he published in the scientific literature. These kinds of data were much in demand by planners of geographical expeditions.

Jane Strachey had begun to use her social position and her husband's connections to further her political interests. She spent a lot of time with prominent scientists such as Joseph Hooker, John Tyndall and John Lubbock, making her 'convinced of the superiority of science in developing superior and charming men'.[11] Joseph Hooker, the director of Kew Gardens, became an especially close family friend, and the two couples travelled together in the United States for a few months in 1872.

By then, Jane had persuaded Richard to spend more time in England for the sake of their still-growing family. This meant he could spend more time with his scientific pursuits. The drawing room of the home in Lancaster Gate became the centre of their new London life. Not only was Jane Strachey an eager intellectual but she was also an energetic socialite as she later confessed to Virginia Stephen: 'My ideal life would be to live entirely in boarding houses.'[12] Later, Leonard Woolf remembered that Sir Richard Strachey was 'a symbol of all that was good and wise and dignified and beautiful, sitting beside his wife at one end of the drawing room, while at the other end, around the piano, there was a noisy hurly burly'.

Jane continued to be especially concerned to help scientists and their ambitions. Not surprisingly these lay in the feminist movement that was just beginning to grow, along with all the other opportunities of those advancing times. She was more intellectually adventurous than Julia Stephen and took an interest in the Fellowship of the New Life, a group campaigning for a wholesome lifestyle and more aware-ness of human sexuality. This was when the German psychiatrist Kraft-Ebbing was writing explicitly about aberrant sexual behaviour and when Havelock Ellis promoted talk in transgender psychology, narcissism and autoerotic practices.

Although there was still much taboo about these topics, biologists such as Ray Lankester were promoting enough anatomical and physio-logical detail from the new investigations of mammalian life cycles that such talk began to remove many of the old stigmas that were around sexuality in urban society. However, it was to take time, and much more education was needed especially in poor communities where the prob-lems were compounded with so many other difficulties. The scientists who were lined up at Darwin's funeral had only a small idea of just how bad some of these living conditions were. Only slowly did they face up to the responsibility of educating the masses about human biology and its bearing on public health and poverty. Instead of focusing on these problems in depth, they gave a lot of attention to another related prob-lem: that of insanity and mental illness.

This was partly because Darwin had encouraged biologists to share his interest in problems of abuse such as slavery and insanity. There were growing numbers of social reformers, and they gave these prob-lems enough publicity for them all to become serious political issues. The sociologists and scientists went on to predict that mental illness would become more widespread because these patients usually had more chil-dren than the mentally well. This encouraged more popular talk and that generated fear across late Victorian society. Together with critical reactions from the popular impressions left by Darwin's writing about human racial diversity, biologists suddenly found themselves having to defend complex charges that were both difficult to explain and probably unfounded.

***

Another of the gentlemen scientists at Darwin's funeral was Francis Galton, then sixty years old and a prominent member of the Royal Geographical Society (RGS). He was a good friend of Richard Strachey, and they would often walk together in Hyde Park and campaign for new explorations within the Empire. Galton was twenty-two years old when his father died in 1844, leaving him rich financially but wounded emotionally. Since then he had oscillated between genius and insanity, though his mental instability forced him to pull out of his finals exams at Cambridge in 1843. One biographer called his anxiety 'an impossible attempt to combine those studies with a very wide range of other interests and occupations'.[13]

In the 1850s, he explored Africa, usually on official projects organised by the RGS. Twenty years later, in 1872, he was president of the geography section of the British Association for the Advancement of Science and was involved with Mr Stanley's search for Dr Livingstone. Little is recorded of Galton's inner feelings that were generated when he met individuals from these remote communities, unlike Darwin's moving accounts of his confrontations with the people in South America, but Galton's own lifestyle was hard, and he looked very stiff for the family portraits. He posed as the typical cold and remote English gentleman.

Lankester recognised Galton as one of the greatest living polymaths. Through a meticulous process of collecting and measuring data, Galton had come to classify fingerprints, weather patterns and many aspects of psychology. He was never short of ideas, and his private income allowed him to be free of targets and the plans of others. In contrast, Lankester was always short of money and hated the humiliation that this often caused; he was especially bitter about his dependence on institutions. It's not clear what he really thought of Galton's position as a wealthy man, never needing to apply for a job and being able to flit from one interest to another. Galton had always quietly competed with his cousin Charles Darwin. Now that his great relative had passed away, he was challenged to take the theory of natural selection a stage further, and he may have seen it as his duty.

Galton's obsession with measurements had convinced him that humans should be monitored regularly to improve our knowledge and care for health. In the 1880s, he set up a portable laboratory in Kensington to measure the physical and behavioural features of visitors to the nearby International Health Exhibition. Lankester saw the queues of expectant volunteers lining up outside Galton's caravan when he walked across Hyde Park, and every time he grew more sceptical that Galton was going to get anything useful from all the effort. However, Galton was one step ahead, having already found a way to show how the workings of the mammalian mind might fit with evolutionary mechanisms. To begin his analyses, he used the data from his Anthropometric Laboratory and began to test his first statistical methods. He plotted out in different ways the height measurements of thousands of parents and their children and found three things: first, taller parents tended to have taller children; second, the children were rarely taller than the parents; and third, shorter parents had only slighter shorter children. He concluded that there was a tendency to revert back towards the average.

Francis Galton married Louisa Butler in 1853, and they moved to 42 Rutland Gate, on the south side of Kensington Gardens, where they lived for the rest of their lives. Without the cost of raising children, they were able to entertain lavishly. In middle-class Victorian tradition, the Galtons would spend every summer 'on tour', mostly in Europe but sometimes in the Lake District or even the United States. These were trips with a scientific or cultural aim, and sometimes they went with friends such as the Stracheys or the Hookers.

Galton also collected data from many of his friends, including the Stephen and Strachey families, from which he concluded that men of the gentry were more likely to have bright children. Although he did not discount environmental influences altogether, he contended that eminence was more likely to be inherited. The work meant a good deal to the members of both families, and many years later Virginia Woolf wrote of it when accounting for her own genius, as well as her mental suffering.

Figure 1.2   Francis Galton

The data were plotted in a way that produced a single bell curve. Galton had a craving for measured objectivity, and the simple shape of the curve interested him because it was based on all the data and showed all its variation together. However, Stephen and Lankester argued with Galton. They accused him of trying to reduce a chaotic system to a highly ordered set of targets. In response, Galton said that this was just the kind of open-minded description and comparative interpretation that they all cherished, even if it gave them something unexpected.

Galton had been in India with Strachey in 1846, at a time when civil unrest against the British made it harder for engineering projects to be completed on time. These events led to the Gwalior Rebellion in 1857 and became a major subject of debate in London, stimulating discussion about how earlier catastrophic events might have shaped history. If such catastrophes had been influential, Lankester asked, was the evolutionary history of other species also stimulated by big shocks? So it became a question to debate: whether catastrophic events play a role both in human history and in evolutionary history? Darwin argued with his cousin Francis Galton about whether evolution progressed slowly or in such sudden bursts, and Darwin came down firmly in favour of very slow changes. Huxley warned Darwin away from such a conclusion. In his 1869 book *Hereditary Genius* Galton presented a theory of sudden change. He compared evolution to the movements of one of his 'toys', a rough stone rolling on a smooth surface: with gentle pressure it moved slightly on the lowest face but when it was suddenly pushed the stone flipped to another face and onto a new position. The slight movements represented Darwin's slow evolutionary changes while the sudden pushes were Galton's big events. This model fuelled Lankester's enthusiasm for catastrophic change in evolution, ideas that he later discussed with Karl Marx.

Not wanting to be left out of such discussions about scientific objectivity and social trends, Richard Strachey wrote an essay for the *Encyclopaedia Britannica* about human migration into Asia. Darwin and Huxley had both helped him understand the link between geography and human racial difference. Strachey used the hereditary patterns of four major groups, Mongolian, Caucasian, Australoid and Negroid, to show that humans had evolved along the same mammalian line.

Interpretations like these were often tested out in informal conversations or lectures given at the RGS in Savile Row. Richard Strachey and Francis Galton were senior members of the society. They both agreed to exert as strong a scientific influence on the society as they could. Galton's credibility came from his early travels in Africa and his desire

Figure 1.3   Richard Strachey

to use science to raise living standards there. In Africa there were plenty of opportunities to measure changing climate and environment. Galton understood the usefulness of such data. As it turned out, it was his attempt to forecast the weather that became his most important achievement, but he also played a major part in getting geography recognised as a subject for study in its own right.

Galton wanted to reduce complex systems of biology to small and simple explanations. Around the time of Darwin's death, Strachey and Galton argued that, since living systems worked by organic chemistry and the laws of physics, biologists needed to reduce their scale of observation to that of chemistry. Galton had already helped Darwin in a fruitless search for the agent of inheritance in the 1870s, when they had bred rabbits and looked in the offspring's blood for signs of inheritance by blending. Now he was on firmer ground: he wanted to find parts of biology that could be measured, to add quantities to the existing descriptions of quality before any further biological investigations could be seen as complete.

At the time, one of the main obstacles to progress in understanding evolution was the absence of data. Many features and variables were recognised, and their functions were partially understood, but there were no measurements of how they changed while working. In March 1883, Galton wrote in the magazine *Fortnightly Review* calling for his Anthropometric Laboratory to be able to gather data from measurements of these features so that a new chapter in understanding life processes could begin. He also wrote a pamphlet called *Inquiries into Human Faculty and Its Development* about the kinds of measurements he was expecting to make and how they might say something about the relationships between anatomy, intelligence, personality, race and culture. He wanted to show how these 'eugenic questions' might relate to what was 'good in stock, [and] hereditarily endowed with noble qualities'.[14] It was his early summary of the project that was to follow.

With yet more invention, and together with Richard Strachey, Galton also developed a strong interest in the ways in which scientific instruments, from microscopes to potentiometers, were used. Therefore, by 1895, scientific instruments were standardised by the Queen's Observatory at Kew. Galton and Strachey wanted to bring this precision into the measurement of animals and plants. Their work at Kew led in 1884 to Strachey representing Britain at the International Prime Meridian Conference in Washington, DC. Strachey succeeded in gaining worldwide acceptance of the Greenwich Observatory as the standard site. The global scale of longitude was added to the credits he shared with Galton and the Meteorological Committee of the Royal Society.

Galton's attempts to measure hereditary trends were the beginning of the study of genetics and cell biology, which, as a result of finer and finer measurement, was soon to shed new light on biology and evolution. Still, without proper laboratories for experiments, the first geneticists faced difficult obstacles. Only a small number of people understood the work. Meanwhile, Galton and Strachey became surrounded by more conventional biologists who worked as explorers and observers, microscopists and anatomists. They were usually working together towards a complete picture of biology, showing its diversity and its environment.

Galton knew he was in a strong position to add to Darwin's groundbreaking theories. He suspected that, as Darwin's cousin, they had both inherited special talent from their brilliant grandfather Erasmus Darwin. Galton was pleased to have the support of the still-influential Thomas Huxley on the important matter of whether evolutionary change occurred slowly over very many generations or suddenly in a single event. He gave a lot of space in his 1869 book to that metaphor of the rough stone needing to be pushed over from one side to another: 'evolution by jerks' as some have put it.

The eccentric lifestyles of Galton, the Stracheys and the Stephens were becoming unsustainable, both economically and intellectually. Their first-class education, their sophistication and their experience of travelling gave them wisdom that they wanted to use responsibly, but this knowledge also made them powerful, and the more they felt threatened by young people of the new generation the more reluctant they became to share that power or give it away. With the appointment of more and more biology graduates to specialist jobs, Lankester was pleased to see that change was on its way. More advances in science and the applications in homes, industry and transport were evident in everyday life, and with better education the rapid changes would continue. There were exciting times ahead for the next generation of scientists, and they were moving from the mansions of Kensington, across Hyde Park, to the cosmopolitan squares of Bloomsbury.

2
Lankester takes over, 1884–90

Ray Lankester was Jodrell Professor of Zoology at UCL from 1874 to 1890 while he also lectured at the associated teaching hospital across Gower Street, University College Hospital. He must have wondered whether, if Darwin were to have a clear successor, this was where he was going to be found. And, if anyone of his own generation were to deserve a funeral like Darwin's, would it be Galton or Stephen, or even Strachey? Would he and his friends be gathering to hear eulogies about one of these men in thirty years' time? And what about himself? How many people would be at *his* funeral? What had he achieved so far? What had he discovered? And had he changed hearts and minds of the public and the scientific community in the way Darwin had?

Lankester had been born into a different kind of family to those of the gentlemen who had attended Darwin's funeral. His father, Edwin Lankester, was a self-made man who had been a doctor with an interest in microscopy and biology. He was an intellectual who had been one of the first professional medical-scientists. He had worked as both a general practitioner and an amateur zoologist, had led the biology section of the British Association for the Advancement of Science for many years and had edited the *Quarterly Journal of Microscopical Science*. The Lankesters' home in Savile Row was not far from the rooms of many of science's learned societies, and Ray often used to call in to the Linnean Society on his way home from school. His parents had named all three boys after famous English biologists – John Ray, Sir Richard Owen and Edward Forbes – and they encouraged their sons to pursue the same interests as their namesakes. Edwin was well known for his campaigns to improve public health in big cities, but, in the 1860s, he got into debt and faced public humiliation, an unfortunate legacy for his three ambitious and clever sons.

Ray showed considerable academic promise and was closely overseen by his enthusiastic father. All the same, the young man's first two years studying science at Cambridge were not happy. His parents thought it was because he was not challenged enough, so they moved him to Oxford. There, he was offered a wider range of subjects and a chance to stay on to specialise in marine biology, but he began to feel the same disappointment as he did in Cambridge. Ray had learnt to dislike the elitism of both Cambridge and Oxford. He was shocked by the lack of interest in science shown by most staff and students, and he became more depressed at Oxford, contrasting it to the challenges he had grown used to at home in London. He was contemptuous of those he called the 'flunkeys, snobs, spendthrifts and social bullies', the students who wanted only to have a good time in their three years there.[1]

However, there was another side to Ray, one that drove him to take advantage of this comparison between Oxford and London and to promote his own beliefs wherever he was. Two good chances, unrelated to his time at Oxford, came his way when he returned home to London. One was from his role model, T. H. Huxley, who taught at the Normal School of Science in Prince Consort Road. In 1870, Lankester was asked to help with the teaching in the new laboratories there. He did this very well and gained high praise from his colleagues and the students. The relationship between Lankester and Huxley flourished. Ray could see something of his own father in Huxley. Both were self-made men who had lived through hard times as well as good ones. It meant they understood one another and shared views about religion, politics and the importance of educational opportunities. Most of all, they had the same confidence in how science was improving the living standards of ordinary people and reducing a lot of the poverty.

Then came the second chance. Anton Dohrn, a wealthy sugar importer interested in promoting Darwin's ideas, was so impressed by the work Lankester and Huxley were doing that he asked them to help set up the world's first real marine biology station in Naples. Nowadays, aspiring scientists register as Ph.D. students, and in Lankester's time they also tended to devote several years to a single major project. The aim of such an experience was to build knowledge and confidence and to demonstrate to the scientific community what the individual was capable of achieving. Often it also led to the publication of papers or a book, and thereby the making of a scientific reputation. As Darwin had worked on barnacles for four years, Lankester chose marine invertebrates and, in particular, the molluscs that grew profusely off the coast of Naples. He worked at Dohrn's laboratory in Naples and became an expert on *Amphioxus*, cuttlefish and several exotic worms. He was also fascinated by the physiology of the electric eel.

While he was writing up this work, Lankester won a resident fellowship back at Oxford. He took a cynical view of the role: 'Expenses much greater than anywhere else, society stale, old broken-down, one-sided pedagogues or pedants, dwelling rooms wretched, stuffy, sickening. No liberty of action, a tyrannical majority which shows its opinion of you, even if you stay away from church.'[2] With such strong views, it is unsurprising that university reform became an important cause for Lankester.

As well as wanting to do away with the stuffy Victorian traditions that many people at the university expected, such as ignoring science and science students, Lankester also had a positive agenda. He wanted to create more opportunities for working men and women to get into university, to develop more practical and open ways of teaching and to apply the study of biology to real social problems. In addition, he wanted universities to be independent of the church and the state and for the colleges to have their own government. This reforming zeal made him unpopular in some parts of the establishment, especially at Oxford, where there were still many conservative, inward-looking clerics. Lankester wanted to replace these clerics with external appointees. He also wanted to have open entrance exams for student places, to establish a student budget for travel and research, to relax the rules of residence in colleges, to prohibit donations of university money to religious institutions and to abolish religious tests. He argued in particular against five of the things the clerics taught the students:

1. The dread of the Unseen God, who will punish us, if not now, then after death.
2. The idea that all physical enjoyment is degrading and will bring out vengeance from the Creator.
3. The judgement that copulation is wrong.
4. The opinion of women as inferior creatures, used by the Devil to bring men to misery.
5. The view that there is a God who can be 'got round'.

Despite being open about their beliefs, Stephen, Huxley, Strachey and Lankester retained their own peculiar brand of dogmatism throughout their lives. It was a dogmatism that was deeply ingrained within contemporary Victorian culture; it was what remained once the Holy Spirit had evaporated.

For some years Lankester had looked for patterns in animal morphology that might show how species were related. Darwin had written to encourage him: 'What ground work you did at Naples! I can clearly see that you will some day become our first star in Natural History.'[3] Huxley added his approval, and, when Robert Grant (the first Lamarckian professor of zoology at UCL and early mentor of Darwin) died in August 1874, Huxley strongly supported Lankester's application to the professorship, though there was no doubt by then that he would get the job.

UCL was just the right place for Ray Lankester's style, his charismatic teaching, his free thinking and his faith in science and a new style of living. The Godless Institution of Gower Street had become well known for attracting liberals rebelling at the social deprivation in the urban slums. They wanted to extend education and culture to the general public. Ever critical of what he had experienced at Oxford, Bloomsbury was the ideal place from which to mount his campaign to propagate and develop Darwin's new model for biology and the ascent of humanity. Not only did that model need more evidence to gain universal acceptance by the scientific community but it also needed responsible representation in the many religious and social organisations that were influenced by its message. Lankester was an obvious spokesperson to take up these challenges. Above all, it was scientific knowledge and method that guided Lankester's professional thinking, and it influenced his management style. The other pleasures of life came second, and he suffered for that, as a lonely bachelor.

Throughout the last decades of the nineteenth century, Bloomsbury was a dirty place, and not many middle-class people wanted to live or work there. The smoke from the new railway stations along Euston Road spread dirt and smells, and unsavoury travellers from the north filled the surrounding hotels. Although there were at least a dozen leafy squares and many streets of grand houses, the Bedford Estate also had a large stock of smaller properties that were leased as flats or boarding houses. There was a workhouse at Mount Pleasant and several slums scattered among the beautiful Georgian architecture. Bloomsbury was known not as a residential area but for its hospitals, the British Museum, a few shops and offices and UCL.

Lankester continued to hold up the traditions of descriptive biology and gave special attention to using well-labelled careful drawings of the specimens. For that substantial record, and for his innovations in marine biology, in 1874, just before his appointment to the chair at UCL, he was elected to the Royal Society. With these early successes, Lankester's destiny was clear. The 27-year-old did not hesitate to promote science as the new requisite for modern society. He had proved to himself and others that he was credible, and he soon became one of the most popular speakers in London. He was attractive to his students and the general public because he was not the member of an elite but one of the growing number of middle-class men who were succeeding on account of their talent and drive rather than their status and connections. They admired his honesty and his refusal to exploit his Oxford connections.

Over the next few years, he built the department in Bloomsbury into what was said by his university to be by far the most active school of zoology in Britain, Cambridge being the only possible competitor. Lankester trained a great series of zoologists who filled very many of the chairs in that subject both at home and in the dominions, and he thus influenced the whole course of zoology in the British Empire. Innovations included starting a museum, organising practical classes with microscope work and dissections, making drawings on large canvasses and even using coloured chalks for blackboard diagrams.

Lankester was one of the best descriptive biologists of his generation. His critical temperament forced him to observe closely and thoughtfully, and he took trouble to see the same feature from as many perspectives as possible. He was also a fastidious man and enjoyed public arguments about science, the nature of truth and the rights of man. Befriending characters as different as Marx and Darwin was very much his style, and he enjoyed playing the social rogue. He went much further than most other biologists in promoting and extending Darwin's scientific and social legacy. Inevitably, this meant that some people were going to disagree with him, and he faced up to this with dignity and humour. However, he did not make the kind of discoveries or other achievements that would be remembered by future generations. Some of his friends, rather than scientists, thought that this was because he was living between two major periods of thought in biology and didn't fit easily into either. The African novelist and political activist, Olive Schreiner, said he was 'the most powerful human being I ever came into contact with; he is like those winged beasts from Nineveh at the British Museum. What you feel is just immense force.'[4] She admired him for challenging many of the doctrines that their own mentors accepted without question, such as traditional gender roles and the growing divide between the rich and the poor.

Ray Lankester may have been a colourful individualist, but he was also an autocratic Victorian professor who was revered and respected by the numerous colleagues and assistants who surrounded him. That kind of relationship set the tone for the many Bloomsbury scientists who had been students of other UCL professors. At the time of Ray Lankester, these included the botanist Daniel Oliver, the physiologist Sir John Sanderson, the chemist Sir William Ramsay and the statistician Karl Pearson.

These men were among the first professional scientists and were authoritarian leaders in their disciplines, selected for that talent, and for their fitting background. Inevitably they were strong personalities though they did have a variety of political and religious affiliations. They were determined to cling to power and expected loyalty from their subordinates. In return, the professors were expected to teach and to 'make new knowledge' in their field, and many of them became strong academic leaders. They were assisted in the teaching by teams of demonstrators, new graduates who helped with the practical classes and gave tutorials. There were also attendants in ginger-coloured coats who displayed apparatus, specimens and blackboard drawings in the lecture theatres. These assistants echoed the professors' interpretations of the subject matter and pressed each student with explanations and details. From them all, rigour and hard work were expected to lead the students in critical and original thinking so that they could understand and make new discoveries by the end of the course. This same approach was as relevant in the arts as it was in science. Very similar procedures were followed in the Slade School of Art, in the medical faculty and in the science faculty. Most of the students were just as original and holistic as one another, and the similarities were to continue for another generation. Indeed, for many of the scientists, writers and artists of Bloomsbury, there was little methodological difference and their similar way of thinking gave them a special unity.

Lankester's strength of personality and broad interests enabled him to be accepted as a scientist just before mainstream academia had acknowledged the existence of biology as a distinct discipline. By 1882 he had achieved much to ensure that his holistic approach was recognised. In Germany he was well known as an English thinker: a leading marine biologist, a firm supporter of Darwin's theories and a talented observer of the structures and lifestyles of invertebrates. He had published a great deal about these species, had offered new explanations of what we now call parallel evolution and had developed new methods for examination by the microscope, enabling observations that were sharp and critical. No wonder he had been admired by none other than Darwin himself.

There were not many professional biologists of Lankester's generation, and most had been nurtured, like him, by the encouragement of their enthusiastic parents. There were also not that many young people who wanted to become biologists because most had been brought up to believe in the literal truth of Genesis and moved well outside Darwin's small circle. Biology had not long been established as a university or school subject, even in those institutions where the physical sciences were on the syllabus. The small number of schools and colleges that did teach biology did it to train doctors in anatomy and how to observe, but things were about to change.

In 1881, Lankester arranged for two new graduate students, Raphael Weldon, a Cambridge zoologist, and Karl Pearson, a mathematician, to go to the Marine Biology Station in Naples to begin a collaborative project. In addition to the usual observations and descriptions, they began to use measurements to distinguish species. One of their projects was to compare the size of eleven different organs from hundreds of specimens of shore crabs. The variations within these measurements gave a normal bell-curve variation, except for those from one feature: the frontal breadth of the carapaces. They used this unusual variation to distinguishing between races of a single species, *Carcinus moenas*. Weldon ended their joint article, 'It cannot be too strongly urged that the problem of animal evolution is essentially a statistical one.'[5] Lankester was furious to hear such a one-sided and opposing view of his own strong beliefs that biology should be seen from as many perspectives as possible. Accurate observations and interpretations of anatomy and life history were the established methods of zoology. Here were two young upstarts giving all their trust in the completely different discipline of mathematics to solve the familiar problems of evolutionary biology. It was an early sign that a split was occurring in how to understand evolution, a split between progressive scientists and those reluctant or too frightened to change.

***

These young scientists, Weldon and Pearson, had fought their way through home, school and college to become professional biologists. Quantification was the new currency, examinations were the new tool in education, and the many big unanswered questions in biology gave plenty of opportunity for prizes. Due to the continuing lack of any new evidence to prove Darwin's theory beyond doubt, enthusiasm for his ideas was declining in popularity through the two decades after his death. That meant many scientists looked around for some breakthrough in the many alternative explanations of life. And, try as he did to make breakthroughs of his own, Lankester became frustrated with his failure to find useful evidence from his classical observations of new species. He often lost his temper with others who were equally unsuccessful in their own work in other spheres. Still, he was determined to find the elusive agents of heredity, which would uphold the theories advanced by Charles Darwin and Alfred Wallace, and there were still plenty of places to seek them out. Lankester was attracting a lot of attention with his studies in embryology, and there were many other biologists investigating new aspects of biology. Consequently, he remained optimistic that clues for the agents of heredity would be found soon.

One reason for this optimism was the sudden birth of many new disciplines of closely related knowledge. At UCL these were greeted with the appointment of budding specialists in the new fields. There was psychology, statistics, biochemistry and physiology, and most of these merged into the programmes of research and teaching that already existed. Lankester excelled in spotting bright young students who were interested in these new subjects, and he encouraged them in the hope of unlocking new clues about the mechanisms of evolution. This was Lankester's expectation in appointing his two Naples students, Karl Pearson and Raphael Weldon, to jobs at UCL, in 1883 and 1889 respectively, and both pioneered statistical methods to analyse the data that Lankester had accumulated. However, there were others who feared the impact of these numerical trends, not knowing where science was taking their ethical values and beliefs.

Some of these cautious observers, such as a young journalist from Ireland called George Bernard Shaw, even shifted their interest back to Lamarck's earlier theory that adaptation to new environments was fixed rather than competitive. One of the biggest embarrassments that continued to haunt scientists was that they could not say whether inheritance was a result of biology or learning. The knowledge vacuum encouraged alternative ways of understanding biological complexity, but Lankester and many others were frustrated. More of the people who had initially hoped that *On the Origin of Species* might at last shift rigid social conventions were becoming frustrated by the lack of direct evidence for the cause of inheritance and found it difficult to continue their support for Darwin. Instead, for them, the work that Lamarck had published back in 1809 gave an adequate solution. His explanation of evolution also allowed them to keep their faith in God, but religious attitudes to science were changing, and there was no going back to the explanations of life that were popular at the beginning of the nineteenth century.

One of the strongest spokespersons for this progress in thinking about evolution had been Lankester's first boss at the Normal School of Science, Thomas Huxley. Huxley's own father had taught mathematics at Ealing School. In 1835, the school was forced to close, and in the absence of an affordable alternative he began to educate his children at home. Among the works that he set the ten-year-old Thomas to read were Thomas Carlyle's defence of slavery, James Hutton's work on geology and even the original Aristotle. When Thomas was just sixteen years old he moved to Bloomsbury and attended Sydenham College of Anatomy in Sussex Street, just behind University College London Hospital. It was very unusual for an unqualified boy to be admitted to such a college to study, but at that time Sydenham College was opening up to new ideas from Jeremy Bentham and the open-minded Scottish poet Thomas Campbell. Huxley did well and was sent to attend advanced courses at Charing Cross Hospital. Through these years his lowly social status meant he had a difficult time as a student, having to fight harder than all the young gentlemen from Oxbridge to secure a place in the London hierarchy of scientists. Perhaps because of this, he was not content to follow the normal medical career pathway. Instead of opting for more expensive training, he applied to the Royal Navy to become a surgeon's mate. In 1846, he was posted to HMS *Rattlesnake* for its exploration of Australasia. Just ten years later he had made the big leap to become professor of natural history at the Normal School of Science, in South Kensington.

Figure 2.1   Thomas Henry Huxley

With such a background, Huxley was all too aware of the traditions that English society required in order to be accepted as a 'gentleman', such as ownership of land or industry, or an Oxbridge education. Many such privileged young men went to work for the civil service or the church. Business was becoming acceptable if it was in the family. Banking or the Stock Exchange would do if you had money. Science did not fit into the picture. For this reason, Huxley found himself on the edge of respectability, a position that made his relationships with an investor such as Charles Darwin and a landed gentleman like Charles Lyell important. Huxley's leadership role at the Royal School of Mines was important to Darwin and Lyell because it meant their work was being taught to future generations. In 1870, one of Huxley's teaching assistants there was Ray Lankester.

In 1882, a few months after Darwin's funeral, the professor of poetry at Cambridge, Matthew Arnold, was invited to give the annual Rede Lecture at the Senate House of Cambridge University. 'Literature and Science' was the name Arnold gave to his talk. Aware of the schism that Huxley and Lankester were trying to bridge, Arnold presented a strategy for healing. He did this simply by changing a few definitions. He proposed that 'literature' should include work by scientists. He argued that Darwin's *On the Origin of Species* and Newton's *The Principia* were just as valuable as literary works as the novels by Eliot or Dickens, while 'science' was just as much a part of systematic knowledge, or *Wissenschaft*, as the study of history or languages.

But Weldon and Pearson's attempt to use measurements to help study evolution continued to be seen as a split from Lankester's structural approach and also opposed Arnold's observation that literature and science worked together. Already the row between Lankester's value on subjective observation, and Galton's attention to objectivity and measurement, was encouraging other divisions to show up. The split that worried the three professors, Arnold, Huxley and Lankester, had hardly appeared in the early 1880s, but it was one that was going to separate western culture later in the next century. Science was becoming associated with the new lower middle class while literature largely remained as the preserve of the Oxbridge-educated elite.

During these years there were many changes across the whole of English society, and no one could say they were not just a little afraid of letting go of the reliable old props in their lives. In such times of stress, many animal populations come together as shoals or herds, better to protect one another and to warn off the enemy. In 1890s England, with its extreme social diversity, many like-minded people also came together in groups.

There were the gentry, the social reformers, and there was this new unit of scientists and artists around UCL, gently influencing one another. They would argue internally but support one another when threatened from outside, and they would be defensive against the unknown. Through the new technology, and their plans to explore beyond boundaries, they were working out the meaning of this outlook on the world.

In England such theories had to face up to a social hierarchy and an associated snobbery, and it showed up in the attitudes of the players on this Bloomsbury stage. The insiders were the Kensington families with allegiance to Oxbridge and who were appointed by patrony. They were about to be challenged by enthusiastic young professional outsiders, men and women who were selected competitively on their own merit. The new theatre was in Bloomsbury, and the director was Ray Lankester.

However, there was to be more to these differences than gentle or even serious rivalry between the gentry and the ordinary people. It was the gentry who were being threatened. So, also, were most of the public, who were threatened and confused by the scientifically inspired changes to their way of life. It was appropriate that biological theory was to offer a way through the idea of planning human populations, using the new knowledge from biology to control human breeding at sustainable levels. Arnold had made his objection to the class system very plain in his lecture. So had Huxley in a speech two years earlier. At the opening of Mason College in Birmingham in 1880, he had stressed the importance of science education for commerce and manufacturing, and argued that physics and chemistry were 'handicrafts' to be learnt by people from all social classes. Significantly, hardly any science was taught then at Oxford or Cambridge, and only a small amount of history and classics at Mason College.

Lankester was one of the few people in a position to understand the relationship between science, population growth and the kind of education available. In 1890, most people could only see the tip of this complex iceberg: the content and extent of science education, how it related to knowledge about religion, wealth, health and sexuality. If humans really were able to control the breeding within their populations, how would they control their attitude to race and mental health? Performances of this production, the use of biological knowledge to control the breeding of different kinds of humans, were only just beginning.

***

A couple of years before Huxley's visit to Birmingham, another very different, but equally disturbing, disagreement took place between Lankester and another group: people who called into question the impact of science on the meaning of life. The group had rooms in Bloomsbury at 8 Upper Bedford Place where they practised spirit-writing on a slate at a guinea a séance. The group's leader was an American called Henry Slade, and Lankester's public argument with the group was to make him famous.

It began with a remarkable address given to the British Association in 1876 and 'sponsored' by Alfred Wallace who, seventeen years earlier, had proposed evolution by natural selection at the same time as Darwin. Now, in front of an audience of leading biologists, Wallace spoke in support of spiritualism as a scientifically valid process. That evening, Lankester wrote a letter to *The Times* complaining that the lecture had brought the association and science into disrepute.

There was much that angered Lankester about spiritualism, a pursuit that became popular in Britain and North America during the 1870s and 1880s, one of the many reactions to Darwin's challenges about the origin of life. The idea of contacting the dead and communicating with them presented scientists with an interesting challenge. In their sessions, transparent observation was forbidden and so tests with technology were not to be taken seriously. However, it was a serious activity for millions of people, who sat at cloth-covered tables, in heavily curtained drawing rooms and listened to every sound in the room with suspicion and hope. Lankester set out to falsify any hypotheses that such occasions presented.

The dispute between Lankester and the spiritualists gained much media attention and culminated in Lankester prosecuting one of Wallace's spiritualist friends for fraud. The hearing at Bow Street Magistrate's Court lasted several weeks and attracted huge public interest. Wallace spoke in the defendant's favour, as did the writer Arthur Conan Doyle. The prosecutors called on a professional illusionist to show how the trick could have been performed. Wallace's friend Henry Slade was found guilty under the Vagrancy Act. After an appeal, he was let off on a technicality but had already gone back to the United States. The affair made séances less fashionable, while Lankester became the most famous professor in the country. It did not do Alfred Wallace any good at all.

Wallace spoke at his best to defend his beliefs directly to Lankester. In the process, he quoted Lankester's own friend Ernst Haeckel: 'In the long run, the man with the most perfect understanding, not the man with the best revolver, would triumph. He would bequeath to his offspring the properties of brain that had promoted his victory.'[6] This argument seemed more acceptable to observers like Lankester than the more exotic argument about spirits, and he satisfied himself with the knowledge that the human mind was just as capable of evolving as any other part of our bodies.

Despite such attempts to reconcile their differences, Wallace was unable to regain any scientific respect from these particular biologists. They were unconcerned that he was not a gentleman, or that he had written about natural selection independently of Darwin the gentleman. Rather, it was because at the British Association meeting he had tried to test the untestable in the name of science. Wallace's new critics thought it was blasphemy for such an eminent biologist to give serious attention to the spiritualist cause at a scientific forum. It was an admission of doubt about explaining life's meaning.

By the 1890s, Lankester had developed a focused view about science and, for some, he typified the popular image of a Victorian Englishman who liked to have a sense of purpose, an aim, something tangible to strive for. However, beneath this, there was also something of the romantic about him, and many people who were alienated by his science found a brave openness that was honest and kind. Like the fictional Faust, he had searched for the mysterious power that bound nature into a whole. On the other hand, Faust saw life on earth as some vast transcendental process that changed under its own force.

These complex interests in Lankester's outlook led him to find his own particular perspective of evolution, one that he refined during his frequent visits to Germany and his friend in the zoology department at the University of Jena, Ernst Haeckel. Although both men saw that Catholics were finding it increasingly hard to argue against evolutionary biologists, they continued to be inspired by Goethe's way of looking at biology. He emphasised the importance of analogy and the fact that similar features were often found in different structures. They had often talked of a particular process in embryology that they called recapitulation. They had noticed that a growing embryo appeared to pass through a series of stages corresponding to the evolutionary history of species.

Lankester and many other late Victorian biologists were excited about what they saw when they examined the stages of early development in the embryo. They saw that at a particular stage in development the embryos of creatures as diverse as lizards, birds and elephants look the same. Although the youngest embryos in a related group usually started off by looking different, they grew to look similar, then became different again. It was during the middle phase of similarity that the body plan was laid down; it was a moment of minimal anatomical divergence. Therefore, Lankester and others thought there was some relationship between the way an embryo developed and the way the species evolved;

they began to think (wrongly) that the stages of a developing embryo followed the same path as its evolutionary history.

On some of his trips to Germany, Lankester and Haeckel worked on a group of salamanders whose development remained at the juvenile stage yet were able to breed quite easily. They suggested that these so-called 'axolotls' were degenerate species, that is, forms that could only have evolved from mechanisms within the organism. They proposed that the evolutionary process at work here was different to anything that might come from one of Lamarck's straight pathways. Degeneration took place within the organism, while the processes that Lamarck advanced were stimulated from the outside environment. In Lamarck's scheme, it was unclear how an early embryonic phase might be inherited, let alone how it might evolve by natural selection.

There was pressure on Lankester from all of his colleagues to find evidence for inheritance and to devise experiments to prove how inheritance might enable evolution by natural selection. However, the units of inheritance and the mechanism of their passing from one generation to another were to remain elusive for several more generations. Even when some good evidence for inheritance did appear, for example, from Gregor Mendel's early experiments on plant breeding, it was difficult to put it in any meaningful context. Although Mendel had published his work in his local scientific journal back in 1866, he had never publicised. As we shall see, its importance was not discovered until 1900. The long-anticipated particles of inheritance remained a mystery, hidden in the library of the monastery garden in Brno.

Degeneration: A Chapter in Darwinism was the title of Lankester's 1879 presidential address to the zoology section of the British Association for the Advancement of Science. In this address, which he later turned into a book, Lankester described scores of species with unused organs, domestic ducks with smaller wings, blind cave animals and even rare families of humans with extraordinary musical talents.

Lankester was a man of contradictions. He was not one to be taken in by oversimplifications of the complex systems of biology. Yet he strongly believed that arguments were a necessary part of the way the human intellect worked things out. So, although he disagreed with much of the new criticism of natural selection, he did admire some of those who were attacking him simply because they were speaking their mind.

Lankester was asking himself whether humans could degenerate to a simpler form of society. Evolution was normally assumed to turn the primitive into the advanced and civilised. Instead of seeing evolution as progress, he was arguing the opposite. Could species sometimes return to the form of an earlier generation? Could this explain different racial and environmental characteristics? As well as bringing about progress and reform, maybe evolution could also bring about regression and decline? This idea was pounced upon by those who saw it as a possible explanation of new tendencies in the arts, such as dandyism, naturalism and mysticism. Others used it to explain how they thought some races had evolved 'less' than European ones. They returned to Darwin's *The Descent of Man* and saw different ways of interpreting the relationships between different races.

Lankester joked that he saw 'degeneration' at work in men who inherited a fortune, but his idea backfired when it was taken up by conservatives in order to argue that God had put humans at the apex of complexity. That position implied that humans were in charge of everything below and could control it all. The argument had first been made in 1809, when the French botanist Jean-Baptiste Pierre Antoine de Monet, Chevalier de la Marck, published his theory of evolution by acquired characters. This earlier theory was now used to explain how strong or dominant humans could control lower forms. In the conservatives' scheme, there was no need for anything complicated like competition. The idea of degeneration allowed them to twist Darwin's theories to their own ends.

One reader who enjoyed *Degeneration: A Chapter in Darwinism* was Karl Marx. Marx especially liked Lankester's suggestion that degeneration could apply to humans as well as to other species. In an appendix to Lankester's work, which Marx arranged to be translated into Russian, Marx added:

> in the case of human societies, it is to be supposed that ultimately a degenerate society would be beaten, repressed, and eventually annihilated by other species... The struggle is so close among civilised men that the possibility of a degeneration and permanent rest does not suggest itself. It is exceedingly probable that a community which aimed at degeneration would end in annihilation.[7]

For Lankester, evolution followed Darwin's branched tree-shaped vision of evolution and not that of Lamarck which saw a lot of long and straight ladders. For Darwin, extinction was a very necessary part of the process of evolution.

The unexpected friendship between Marx and Lankester grew out of one of the latter's lectures in the 1870s, in which he asked whether evolutionary features such as degeneration and extinctions occurred at the level of an individual or a whole population. An American archaeologist called Charles Waldstein was in the audience and stayed on to talk to Lankester about the values of socialism. Waldstein was concerned that socialism conflicted with individual liberty. After their discussion, Waldstein introduced Lankester to Karl Marx in order that they could talk about evolution. Both believed, however, that God did not have a hand in these processes; there was not going to be a fixed end or goal. Lankester regarded this as a satisfactory beginning.

Lankester decided to stay on as a poorly paid Bloomsbury professor, and Marx continued to offer him theoretical advice. Eleanor (nicknamed Tussy) often joined in their conversations, and Lankester advised her on the difficulties created by ageing parents with no money. Eventually Lankester arranged for his friend Dr Bryan Donkin to attend Eleanor's ailing parents. The doctor had a practice near Portman Square and also worked to diagnose neurological disorders in prisoners. After Karl Marx's death in 1883, Lankester comforted Eleanor through her own grief and loneliness.

Dr Donkin shared Lankester's belief that science constituted a search for truth. The two friends had been brought up to trust in God; now, however, they only had faith in science. They believed that they owed society a priestly responsibility to test scientifically any mystery they encountered, and they became dogmatic as a result. Their conviction that this view was correct sometimes produced intense antagonism among their colleagues, and especially among the many quacks who worked in late-nineteenth-century medicine. While Lankester was trying to take superstition out of the public understanding of evolution, Donkin was trying to help improve health care in prisons and workhouses. Both men soon realised that their missions upset the establishment and lacked the support of most political organisations.

Lankester was also interested in the idea that an organism's adaptation to its environment might be driven by a natural energy, what became known as l'élan vitale. According to this idea from French philosophy, nature was a single complex system. Lankester was aware that just as Darwin had studied the diversity of barnacles during his voyage on the Beagle, so Marx had seen political unrest in Germany and had studied the Epicurean will against pessimism as part of his PhD thesis. They had all reached a similar conclusion from their very different studies that large systems change within themselves.

Not all of Lankester's judgements were sound, not least his early support for one of his zoology acquaintances, Edward Aveling, a man who said he was fanatical about Marx's political ideas. There was nothing that Lankester liked more than a good argument and talking frankly about politics, and Darwin's ideas was part of the topical diet in these intellectual circles. Aveling was two years Lankester's junior and had been born into a well-off family in Stoke Newington. Through Lankester's influence at UCL, he switched from the medical school to study zoology. He graduated in 1870. As a bright and promising biologist, he was highly recommended to a lectureship at King's College, but his atheism did not endear him to the authorities there, and he was soon asked to leave. Not one to give in, he wrote a popular account of evolution by natural selection that became an important primer for budding biologists. *The Student's Darwin* was published in 1881 and formed the basis of several meetings that Aveling arranged with Darwin about evolution and religion. He based another book, *God Dies, Nature Remains* on these meetings, in which he claimed that Darwin was an atheist. The controversy the book caused meant that Aveling lost those supporters he had left in academic circles. Aveling thus turned to politics and became a leading member of the Secular Society. He soon gained a reputation in the society for claiming excessive expenses and embezzling funds, and in his social circles he became known for borrowing money and not repaying it. That was when Eleanor Marx fell in love with him, beginning a relationship that ended in tragedy.

Figure 2.2   Edward Aveling (1880)

Figure 2.3   Eleanor Marx (1886)

Aware of the difficult challenges facing those who inherited Darwin's legacy, Lankester maintained his belief that science, more than anything else, could help improve humanity. Only through a better understanding of biology would humans tackle the problems of poverty and war. As Ray Lankester matured he could understand just how vast was the challenge ahead for biology. It was not just the supreme majesty connected with the meaning of life, its origin, maintenance and adaptation. For him, there were also the topical problems of poverty, education and corruption. He was asking himself whether the gentlemen scientists could survive, how the new professional scientists could be managed and funded, and how their work could be applied to help humankind in its struggle to survive within the limits of the planet earth. Fortunately for Lankester, the work was interesting and exciting. He was having fun within this group of Bloomsbury scientists.

3
Eccentric campaigners, 1890s

Since 1826, when Thomas Campbell founded UCL, Bloomsbury had been a haunt for bohemians who were attracted to the college, the bookshops, the British Museum and the cheap rooms and hotels. Ray Lankester fitted well into this intellectual world: Darwinism and its associated questions about the place of man in nature interested many of the people who worked or lived there. In Bloomsbury, scientists rubbed shoulders with writers, artists, lawyers and civil servants. There were eccentric young men and women campaigning for their own cause, and many came from humble backgrounds. They had all been strongly influenced one way or the other by the continuing debate about a fairer kind of society and many of them thought that imminent breakthroughs in biology would control the growing population somehow to achieve this.

In the 1890s, there were still no public recruitment procedures for institutions to follow when making new appointments. Usually, jobs were offered privately, to people who knew one another, often in the same family. Most new scientists had studied physics or chemistry and went straight to jobs in industry. Most of those from the London colleges went on to Oxbridge or enrolled as medical students at one of the London teaching hospitals. In many families the sons followed the same profession as their fathers and they went to the same college and hospital.

Among these groups of enthusiastic students were some who had special talents for creativity and original thought, people who felt passionately about some intricacies in living systems and who recognised that biology was at the beginning of some kind of dark and mysterious journey. They were excited to be involved with new experiments and experiences, to dig deeply inside organs and tissues, right into their very small cells that had never been investigated. They were thrilled to be working at this time, when new kinds of processes were being discovered and when new relationships between organisms and species and environments were being recognised. They were uncertain people working with uncertain objects in uncertain times. Inevitably, this meant that the people themselves tended to be rather unusual.

The Bloomsbury intellectuals often worked in the Round Reading Room at the British Museum, in Great Russell Street. It was here that Marx researched his critique of political economy, *Das Kapital*, and where Leslie Stephen drew on the rich collections of the library when he was editor of *The Dictionary of National Biography*. Another frequent visitor was R. L. Stevenson who was developing the character of Long John Silver for his novel, *Treasure Island*. Stevenson based Silver on Bill Henley, a Scottish poet with a wooden leg and one of Ray Lankester's Savile Club friends.

Another of Lankester's friends was the Irish playwright George Bernard Shaw. Lankester and Shaw often attended one another's public talks and were regular dinner companions. The twenty-year-old Shaw had moved from Dublin to Bloomsbury in 1876 and lived with his mother at 29 Fitzroy Square. She gave him a pound a week to work in the Round

Figure 3.1 George Bernard Shaw (1889)

Reading Room where he befriended Eleanor Marx, Edward Aveling and Thomas Hardy. This was before Shaw began to write his plays, and he spent his time preparing popular articles and talks about socialism. He was part of the inner circle of the early Fabians and saw himself as 'a complete outsider' from the English literary scene, swearing that he 'would avoid literary society like a plague. Literary men should never associate with one another.'[1] Instead, he involved himself in a busy programme of public lectures about the social revolution. He developed a witty and humorous style of communication. He used this to speak to thousands of people in Trafalgar Square as well as to small groups of poor people in the East End slums. That is to say, he gave equally passionate and amusing accounts of the new dawn of socialism. After sixty-six public presentations in one year alone, he had found a powerful way of contributing to the political debate of that time.

Shaw believed that evolution must be driven by mind and purpose. He was convinced that Darwin's proposals about natural selection could not work because they made no provision for these life forces. Instead, he said Darwinism led to a moral vacuum. There was 'a hideous fatalism about it [and] nothing natural about an accident'.[2] Consequently, Shaw went back to study Lamarck's ideas of inheritance by acquired characters and spent a lot of his time promoting the early theory. Here was a will to change, to exercise those faculties that were useful and to suppress those that were not. Living organisms changed because they wanted to.

Meanwhile, more and more people were finding themselves stuck in poverty. No one cared for their plight, there was no attempt to improve their bad diet, and their lifestyle meant that diseases were spreading. Despite a good sewage system, cholera was still a killer in Bloomsbury, especially in the hospitals and workhouses. Prostitution and street crime were growing fast and causing fear among the public. Concerned doctors and lawyers were finding it hard to exert any influence on government or to arrange any social benefits; they were just as helpless as were the majority of the victims. Because of its central location, Bloomsbury also accommodated more than its fair share of unwanted children, at the Foundling Hospital and the workhouses in Endell Street.

In 1888, concern for the deteriorating state of public health in the big cities had brought Lankester and Dr Bryan Donkin together at seminars on biology and medicine that were becoming frequent at the Bloomsbury medical schools. Lankester's father had been a London doctor, specialising in problems of public health, and encouraged Ray and his other sons to care for the plight of the underdogs. R. L. Stevenson often talked about this over dinner with Lankester and Donkin, and this may have helped him come up with the characters of Jekyll and Hyde. All three men were preoccupied with evolutionary degeneration, the prospect that the human race might deteriorate into forms resembling their primitive ancestors. Stevenson viewed Victorians as being hypocritical about morality, and he wrote about the extremes of rich and poor in several of his stories. Lankester, meanwhile, lectured about the wide range of behaviour in human races. They both used the same arguments to explain the good and the evil in people like Long John Silver, Jude the Obscure and Dr Jekyll.

During these conversations, around 1885, George Bernard Shaw was writing book reviews for the *Pall Mall Gazette*, a forerunner to the *London Evening Standard*. Despite his earlier vow not to mix socially with the literati, we may imagine him inviting some of his friends to a house party at his home in Fitzroy Square. Here is Lankester walking into the big scruffy house that had changed little since it was built in the 1820s. The room is full of cigarette smoke. People are standing with drinks in their hands, speculating about the cause of the latest East End street murder. Their host holds the attention of a group in the middle of the room; they let out loud shouts and laughter. Lankester has just returned from Germany where he was visiting the laboratory of another distinguished biologist, August Weismann, who is carrying out simple experiments to compare Darwin's theory of natural selection with Lamarck's acquisition by use. Shaw has asked whether they might be relevant to Dr Jekyll's predicament.

Clearly, the Irishman has read his Darwin and has been amused to learn in *On the Origin of Species* that between Dublin and Liverpool, on the Isle of Man, wild cats without tails have come into being. In Germany, apparently, there are packs of tailless foxes as well, which have recently been wiped out by Prince Wilhelm zu Solms-Braunfels' hunting. Are these cats missing tails due to some biological process of degeneration like that set out by Lankester? Could this even explain the mystery that lies at the heart of the Dr Jekyll and Mr Hyde story? Is Dr Jekyll's double like a tailless cat, created in a laboratory using a special potion? Stranger things are happening, the Irishman might have argued.

This might also have been an opportunity for Lankester to expound on Weismann's ongoing yet still unpublished experiments with tails. Several German biologists still preferred Lamarck's theory of evolution to Darwin's, and suspected that animals and plants could transmit mutilated organs. Weismann set about cutting off animals' tails to see if there was any physiological reaction in the offspring. He started with twelve white mice, seven females and five males, which soon bred 901 youngsters over five generations. So far, each one had a tail, and mighty fine ones at that, varying from 10.5 to 12.00 cm in length. Weismann's theory was that inherited characters were transmitted from one generation to the next during sexual reproduction, through interaction of egg and sperm. Non-sexual structural organs played no part in it, and the mouse-tail experiments were supporting his view.

Shaw ridiculed the arrogance of the scientists who were using such naive experiments to test Darwin's and Lamarck's different views of evolution. He thought it was ludicrous to attempt to prove evolution experimentally. It was, he thought, like using logic to prove that God exists.

To make his point, Shaw might have told the two stories that he later included in his preface to the plays *Back to Methuselah*. The first likened Weismann's experiment to one he had performed himself ten years earlier at a party given by a group of American evangelists. At this party, one of the guests, Charles Bradlaugh, the most formidable atheist on the Secularist platform, took out his watch and challenged the Almighty to strike him dead in five minutes if He really existed and disapproved of atheism. Shaw could not resist putting his own watch on the table as well. The room fell silent. 'Only the leader of the evangelical party was a little preoccupied until five minutes had elapsed and the weather was still calm.'[3]

In the other story, Shaw ridiculed a logical explanation of God's existence. One day at the Brompton Oratory he had talked to a priest called Father Addis, and a conversation developed that took the form of two men working a saw: Addis pushed one way, Shaw the other, and between them they cut nothing. 'The universe exists, somebody must have made it', claimed Addis. To which Shaw replied:

> If that somebody exists, somebody must have made him ... I grant that for the sake of argument. I grant you a maker of God. I grant you a maker of the maker of God. I grant you as long a line of makers as you please; but an infinity of makers is unthinkable and extravagant.[4]

Then Shaw offered a benediction: 'By your leave, it is as easy for me to believe that the universe made itself as that a maker of a universe made himself.'[5] For Shaw, science and religion were separate human concepts; one had little to offer the other.

With such party talk, Shaw might have made evolution seem brighter and more palatable than the scientists were able to, endowing it with a benevolent life force. Shaw acquired this idea, quite uncritically, from his friend and adviser, Samuel Butler. Shaw himself went on to develop the popular idea of creative evolution as an impulse of the loftiest kind, an invisible and inexplicable life force for humanity to seek out their utopia. Like Mark Antony, Shaw did not believe that life can be 'shrunk to this little measure' but that was just how scientific reductionism was dealing with evolution. In Stevenson's story of Dr Jekyll, the inward-looking scientist can only provide part of the explanation of life, for there are 'doubles' in conflict everywhere and many intermediates in between.

***

Lankester could see the social divisions between the different communities located in Bloomsbury. The gap between rich and poor was increasing, and it was becoming harder for religion to hold communities together. Although science offered labour-saving technologies and better communications, it could not give clear ethical guidance. Many people found this difficult to accept or used it as a reason to want to limit the realm of influence that scientists could consider. George Bernard Shaw, for example, worried that scientists in general, and Darwinists, in particular, should keep their noses out of the affairs of others. This set him against Lankester for his ambition to use scientific methods to reduce the gap between the extremes of rich and poor, by both raising health standards and improving education. Thomas Huxley had the same sense of duty, but he was getting old and found it difficult to get much done. Lankester, however, was in a strong position to make a difference. He was one of the few people to understand how forces of nature controlled the human predicament, how different social groups supported one another and how their social interactions strengthened their separate identities and kept them going. He was one of the few people to agree with both Galton and Marx: that to bring about change within this stable state there had to be some sudden stimulus from outside the system. Unfortunately, a major difficulty was to know what kind of stimulus was required and how to apply it. Galton was too eccentric and isolated to care and Marx was angry enough to advocate direct violence.

It was obvious to Lankester that the way to stimulate social improvement was through the slow application of science, technology and education. Already, the pace of technological advance was overwhelming to many people, regardless of class, and many institutions began to lose their sense of purpose. So much change had happened over the previous ten years: electric lighting and power, motor cars, telephones, gas heating. The high pace of change convinced many people that urgent improvements in science education were necessary and hard to keep up at the same rate. Universities, colleges and schools had to be streamlined to make them more conducive to technology and science, and teaching had to address real social problems. The challenges were formidable, but Lankester knew what needed to change and he thought he knew how to bring it about.

Lankester was well known as a charismatic teacher who was able to make biology topical. He made grand entries to the lecture theatre at one o'clock sharp, holding the tails of his frockcoat as he climbed to the rostrum. When he reached the lectern, he smiled and gave a little bow; by then the students were standing and burst into warm applause. He spoke with wisdom and humour. An attendant displayed large sheets with anatomical drawings. Test tubes bubbled on the demonstration table. Lantern slides showed photographs of animals in the wild. The hush of the audience was punctuated by regular gasps. Finally the professor thanked the gentlemen for their attention and left with a flourish.

Throughout, high priority was given to the value of human observations, the subjects that a particular biologist could see and go on to interpret in his own particular way. It was accountable and repeatable and enabled an individual to show their feelings. In this way the biologist was more an artist than many cared to admit, certainly more than a chemist interested in distinct molecular structures or a physicist with particular mathematics. This attitude led to a network of like-minded biologists for whom creativity came from observed detail, relating what was seen to some function that became diagnostic of a particular species. New patterns emerged and a careful choice of words could give an unambiguous description of something beautiful. Lankester and his students realised that being able to produce good pictures and prose in this way was a gift that was as necessary for the biologist as for the artist.

Lankester was in high demand for talks and newspaper articles. Like the French comparative anatomist Georges Cuvier, he used drawings in his lectures and articles to interpret biological structures, something that photographs could never do so well. Drawings brought to life a static shape and his lines added rhythm and function to the image. Many of Lankester's vivid drawings survive in his writings, especially in his series of non-specialist articles, 'Science from an Easy Chair'. These drawings are made with confident strokes of black ink on white paper and are clearly derived from their equivalent on the blackboard in the lecture theatre.

Lankester valued conversation as a means by which scientific advances were made. He insisted that all his teaching departments had a laboratory and a museum. Lankester and his colleagues worked in this same space and that allowed dialogue to take place. He put a lot of emphasis on debate and exchange, rather than the acceptance of absolute fact and certainty. He and his friends were challenging orthodoxies by appealing to what people felt about their observations and comparative experiences, and he asked them to express these feelings to one another. It was a demanding approach for teachers and students in the face of a rigid orthodoxy that came from decades of strict religious indoctrination. The mix of people from different backgrounds and interests helped give Bloomsbury a distinctive atmosphere. It was one of the few places where simple divisions between good and evil, right and wrong, were questioned and undermined. This was unsettling to many but satisfying to most.

In 1884, another controversial figure entered the challenging world of Bloomsbury science: the mathematician Karl Pearson. Pearson was twenty-seven years old when he was appointed professor of applied mathematics and mechanics at UCL. His role was to analyse the growing amount of data available from social and biological sources. Pearson had been one of the students whom Lankester invited to the Naples marine biology station a few years earlier. There he had developed exciting new ideas about analysing the range of shapes and sizes in jellyfish and octopuses, working through and making sense of the seemingly chaotic sets of data gathered at Galton's Anthropometric Laboratory. The methods he devised turned into an increasingly large scheme: the completely new discipline of statistics, which became his vocation. He was one of the first to analyse data from biology, and to show social trends. Lankester's quality was being joined by Pearson's quantity, but Pearson's sharp focus on data was not in line with the more flexible intellects of his Bloomsbury colleagues. Tensions continued to develop between the so-called hard and soft approaches.

Pearson was brought up in an archetypal Victorian middle-class family in Mecklenberg Square on the eastern edge of Bloomsbury. His father was a successful barrister who paid no attention to his family during term-time. During the holidays, however, he would take the youngsters shooting and fishing. Karl attended school in London and then studied mathematics at Cambridge, where, in 1879, he received a first and was elected to a fellowship at King's College. He was a natural dissenter and argued with his Cambridge colleagues about the divinity classes that, as a fellow, he had to attend. Much more challenging to his mind was the poverty he remembered from the streets in London, close to his home in Mecklenberg Square.

In Doughty Street, just across the road from the Pearson family home, Charles Dickens had written about Oliver Twist, forty years before, and it was still a rough neighbourhood. The bobbies had just swapped their rattles for whistles, but crime and the threat of violence were still common. The cobbled streets were covered in horse manure, and the air was polluted with smoke from hundreds of coal fires. The Pearsons lived in warm elegance surrounded by servants and possessions, while some of their neighbours lived four to a room in filth and squalor.

Karl Pearson was determined to help change these discrepancies and saw that one way forward was by somehow limiting the patterns of human breeding that were clearly getting out of control. His peculiar brand of mathematics searched for patterns in nature, then in biology, and engineering, and finally in people and politics. He was hopeful that, by working on the group rather than an individual, the families would become smaller. He believed that in nature, organisms lived together in geometrically describable shapes. For him and others in the Bloomsbury culture, science was at the centre of everything: nature, human society,

Figure 3.2   Karl Pearson

feelings, individualism and existence. Science observed and measured these things, and to understand the numbers one had to analyse
the data.

From 1879 to 1884, like his colleague Lankester, Pearson spent a
lot of time in Germany, exploring its literature and philosophy. He was
one of the great polymaths, enjoying both science and the romance of
great literature and music, but he struggled to see his own clear career
path. For a time he followed in his father's footsteps and trained as a
lawyer. However, he also admired Marx's left-wing politics – indeed he
admired them so much that he changed his name from Carl to Karl – but
what excited him most were the trends emerging from his new statistical
mathematics and the data being accumulated by people such as Francis
Galton.

Pearson liked to explore and articulate his ideas in both scientific
and literary forms. For example, when he was twenty-three years old he
wrote a novel called *The New Werther* in which he expressed some of his
passionate feelings:

> What meaning has the word 'kiss' to him who does not know that
> through the electric contact of a moment two fiery souls may feel
> united for an eternity? What meaning has the word 'life' for him
> who has only existed in order to hand down his name to posterity in
> the footnotes of a classic or as inventor of an integral?[6]

Even passion had to be demonstrated scientifically, however hard it was
to quantify. Physical attraction was one of Karl's interests, and he rated
its biological importance higher than any of its more popular emotions,
saying that it 'had other and more worldly elements than mere sexual
passion'.[7]

With Lankester's zoology department on his doorstep, it was inevitable that the energetic young Pearson used his new statistics to analyse the data from studies of sexual reproduction. Simultaneously, his
friend Havelock Ellis was telling him the latest opinions on human sexual
behaviour, which were appearing in his books titled *The Nationalisation
of Health* (1892) and *Man and Woman* (1894). They went together to
meetings of the Fellowship of the New Life, the precursor of the Fabians,
where they also met George Bernard Shaw, Olive Schreiner and Eleanor
Marx. In such a group, Pearson was bound to open up about his own
socialism and explain how it might follow the patterns of evolution. He
wanted to test the idea that evolution worked at the level of the group
rather than the individual.

In 1885, wanting more debate about the implications of sexual relationships between men and women, and with encouragement from two of Lankester's friends, Olive Schreiner and Bryan Donkin, Pearson set up the Men and Women's Club, an informal group of open-minded people interested in the social pressures on relationships between men and women. The group, which had ten to fifteen members, met in one another's homes, most often in Schreiner's rooms in Guilford Street and Pearson's home just round the corner in Mecklenberg Square. In such intimate surroundings, the domestic circumstances began to interfere with the more general purpose of the meetings. Without clear boundaries, their frank discussions of how they felt about making love began to intrude on their relationships with one another, and sure enough, Pearson, Donkin and Schreiner soon became bound up in romantic ties.

Olive Schreiner did not want just to upset apple carts, she wanted to get to the heart of social prejudice about gender and biology. That is what brought her to the Men and Women's Club and drew her to people like Pearson and Donkin. Her *Story of an African Farm* was published in 1883 and received acclaim for the manner in which it tackled some of the urgent issues of its day, issues ranging from agnosticism to the treatment of women. Schreiner met Havelock Ellis at the club in 1884; in their conversations, they compared the results from some of Lankester's embryology research to some of the taboos of human reproduction. The club was one of a number of radical discussion groups to which she belonged and they brought her into contact with many important socialists of the time. Schreiner was insistent on the critical importance of woman's equality and the need to consider men's role as well as women's when looking at gender relationships. Eventually she submitted to the romantic demonstrations of Bryan Donkin but soon tired of his attentions and instead wrote many admiring letters to Karl Pearson. He did not share her enthusiasm and in anger she gave up both relationships. The experience encouraged her to write a novel called *From Man to Man* and a new introduction to Mary Wollstonecraft's *A Vindication of the Rights of Women*. So many of their beliefs were in conflict with the establishment that Sir Leslie Stephen saw to it that neither book was reviewed in the journals which were under his influence.

Pearson's followers were the first in Bloomsbury to use the new biology to make a new social-science 'grammar' for those anxious about the dissolution of individuality. At first thought, this enthusiasm for the Men and Women's Club came from socialists and from feminists like Olive Schreiner. The suffragettes were soon to realise how far there was to go

Figure 3.3   Olive Schreiner

before male–female interactions became equal, and men like Pearson were not going to help their cause directly. Only with such reflections at the club's meetings was the full social significance of these gender differences being realised by the scientists present.

These organisers of the Men and Women's Club prided themselves on the radical nature of the discussions that took place there. Pearson discussed topics such as male–female attraction and suggested that 'evolution implanted in woman a desire for children, just as it has implanted in men a desire for woman'.[8] Olive Schreiner showed interest in these theories as part of her own practice of free love; such theories were a way for her to overcome her earlier shame about enjoying masochism and other sexual practices. Whether this was part of a game to attract Pearson is unclear, but later, when she discussed her experiences with Havelock Ellis, she wondered whether she should take bromide to diminish her sexual urge. Still, Pearson would have none of it, neither the passion nor the chemistry.

Sexual politics had entered the political domain a few decades earlier with debate in Parliament in 1867 and was now reaching another peak. Pearson began to argue that women had a particular role to play in stabilising groups within a whole population, what he called a sexual tie between individual members. According to this argument, politics and government were natural systems that needed women in order to survive and evolve. It was a radical claim that led Pearson to suggest a kindred tie between successive generations. He sensed that this link between sexuality and groups might have political connotations; it might even imply that socialism had a scientific foundation. From such discussions about men and women, Pearson went on to argue that science pointed in a different direction to art. Art existed at the opposite end of the spectrum to science; science revealed patterns made by large groups of atoms or organisms, while art was created by individuals.

These were ambitious theories, typical of Pearson's broad interests in nature and how it worked. Part of his difficulty was finding proofs. He had very little data to feed into his statistics. It was also difficult to know what data were relevant to his claims. There were no easy ways to get rid of the background noise.

***

All across Europe in the 1880s natural scientists were searching for biological keys to the understanding of human evolution, and they based their ideas on Darwin's theory of evolution by natural selection. Darwin's 1871 book, *The Descent of Man*, had set the scene for social evolution but added no fundamental details to the main theory. The book confirmed that all species have a common origin within a single tree of life and that species adapt to changing environments; those species that do not fit in become extinct. As knowledge of human evolution became more complicated, the philosopher and sociologist Herbert Spencer coined the phrase 'survival of the fittest', while Friedrich Nietzsche began to think about the possibility of the emergence of a 'super-human'.

Two other non-scientists, living and working in Bloomsbury through into the 1890s, were regular users of the British Museum reading room, and had a challenging impact on ideas about evolution, biology and society. One was the novelist Samuel Butler, the other a tax collector from Ireland, Benjamin Kidd. Both Butler and Kidd had a good sense of how ordinary people thought about the new scientific explanations of life and, like those, had given Darwin the benefit of the doubt. However, they were both worried that it had been more than twenty-five years since the publication of *On the Origin of Species*, and no clear proof had yet emerged. Many people were impatient for answers. Persistent poverty alone was good enough reason for many to return to religion, and Butler persuaded many leaders in the church to quietly replace Darwin's theory with that of his predecessor Lamarck so as to allow a return to the story of creationism. In different ways, Kidd argued that evolution and religion were so different that they could not be compared, which allowed him to accept both God and Darwin.

Because Butler was reclusive, shunned institutions and hated fashionable society, he set out his thinking in two little-publicised books. Few people read them until after his death. It turned out that Butler also pioneered portrait photography, painted and wrote two novels and many important shorter works. For the second half of his life he lived just to the south of Bloomsbury, in Chancery Lane. This narrow cobbled street served lawyers, the Public Record Office and the Patent Office, all mixed in with newsagents and Georgian houses. Even though Butler had no important interactions with any of the Bloomsbury scientists, the influence of his writings on Darwinism was profound. This reawakened Lamarckian ideas in the minds of George Bernard Shaw and E. M. Forster.

With his lonely confidence, Butler set about looking for a good way forward to resolve the divisions between religion and science. Much to his domineering father's surprise, he had done well as a sheep farmer in New Zealand and had returned to London as an intellectual gentleman of independent means. He became excited about *On the Origin of Species* as a sensible alternative to his father's religious way of life. He also came across St George Mivart's cleverly titled *Genesis of Species*, a book that criticised Darwin's emphasis on competition between populations, which he called 'unrestrained licentiousness'.[9] Mivart, like Butler, favoured 'use-inheritance' as a means to explain adaptation to changing environments. Butler outlined this and other ideas in a book called *Life and Habit*, published in 1877.

In the British Museum reading room Butler was likely to have known Bernard Shaw, and they would have been unable to resist sharing their approval of Lamarckian evolution. Butler's enthusiasm for this led him to the idea, ridiculed by most scientists then, that evolutionary processes would some day enable machines to replace humans. His utopian novel *Erewhon*, published anonymously in 1872, explored this theme. The novel offers a Lamarckian explanation of adaptation in which fitness to a new environment could be acquired in a generation. Later, Butler regretted that: 'reviewers have in some cases been inclined to treat the chapters on Machines as an attempt to reduce Mr Darwin's theory to an absurdity. Nothing could be further from my intention, and few things would be more distasteful to me than any attempt to laugh at Mr Darwin.'[10] If the irony in his novel wasn't meant to evoke laughter, it

Figure 3.4   Samuel Butler

was certainly supposed to elicit a smile, because it put more emphasis on 'adaptations by use' than on natural selection. Nevertheless, such a return to Lamarck's ideas, originally expounded in the early 1800s, upset and disappointed Darwin's supporters. However, Butler's confidence grew, and in 1884 he wrote of Darwin's work: 'To state this doctrine is to arouse instinctive loathing: it is my fortunate task to maintain that such a nightmare of waste and death is as baseless as it is repulsive.'[11] As a non-scientist, Butler saw evolutionary biology from the perspective of ordinary people, and this laid him open to criticism from the scientific establishment, but it also promoted the strong popular move away from Darwin's theories.

Butler was a man who did not fit in easily with any of the London scientists or even artists, but he still made an important mark on how they were seen by the British public. His books were easy for non-scientists to understand and were widely read. They made him a spokesperson for many who opposed Darwin's theories. Butler was concerned that, in the future, science and Darwin's theory, in particular, was going to destroy the very social structures that improved people's lives. That was why he came back from his travels across the Empire with a clear message for the people about how to improve their lot. However, he did not shout it loudly enough for people to hear, and he failed to impress any people in positions of power.

Butler is now best known for *The Way of All Flesh*, a novel about the oppression of family life in Victorian England. In the novel, the main character 'made up for having been a servile son, by becoming a bullying father'.[12] The literary critic Edmund Gosse said that *The Way of All Flesh* was a 'romance founded on a recollection' for it tried to justify the brutality of society in terms of Lamarckian evolution.[13] Although it was finished in 1885, Butler never found the courage to share it with anyone, and the manuscript's publication had to wait until after his death. Perhaps this was just as well because it avoided the public confrontation with Ray Lankester that would surely have erupted. Even so, Lankester surmised what Butler was thinking and in 1884 complained in a letter to the magazine *The Athenaeum* about Butler's mistaken belief in the inheritance of acquired characters. This inspired a heated correspondence in *The Times* and *Nature* as Lamarck's old ideas were revived by many more non-scientists in Britain and America.

Across London in the 1890s, scientists and non-scientists alike were asking, What is nature? What does it mean to be human? Although Darwin remained popular with the establishment, and the circumstantial evidence for his theories was compelling, there was still no clear evidence that he was right. Most people had their own feelings about the meaning of life, and their feelings were summed up most effectively by another reading-room regular, Benjamin Kidd. He was as much an outsider from the scientific establishment as Samuel Butler but with views more in tune with the opinions of Charles Darwin and the British public of the time.

Kidd published his popular reconciliation of religion and evolution in 1894, a book titled *Social Evolution*. It argued that both religion and evolution can work together. Kidd had been brought up in the Irish countryside to love nature, and he moved to London in the 1870s. He was clerk to the Board of Inland Revenue at Somerset House and spent as much free time as possible at the British Museum where he read the classics of natural history. Kidd wrote about nature for popular magazines, and, by 1884, had made quite a name for himself. His nostalgia for County Clare encouraged him to specialise on the formation and biology of peat bogs. He also became an authority on the behaviour of bees and pollination mechanisms. He bought his own microscope from the makers in Mortimer Street and, by 1889, was knowledgeable enough to hold his own with some of the newly professional biologists at the Linnaean Society. A year later, he visited August Weismann in Freiberg to discuss the way insects appeared to discourage plant inbreeding. Kidd also found himself plunged into the debate that was then raging about whether natural selection was at the level of the individual or the group. He was excited that his opinion seemed to matter to professional scientists. It was the first time that he realised just how influential Darwin's work was and how powerful a force in evolution natural selection could be.

Eventually, the hours that Kidd devoted to reading and watching patterns in bees' behaviour paid off and helped him see that these species were living as groups rather than as individuals. He shifted his attention from bees to humans. He explored his peers' devotion to the good of their group and soon came up with a radical idea. Kidd illustrated how groups worked together in the military, and in village and family life. He showed that in each environment and situation a clearly defined goal had to be planned for the team rather than an individual. He then shifted his focus from these human groups to religious utopias, aware that selection needs extinction to move forward: one of his favourite themes was to admit that religion provides for failure. Socialism was a new human system that aimed to satisfy the group. He argued, however, that it could not work because it supported the weak. Kidd approved of Darwin's theory of evolution by natural selection that contained the idea that some had to fail. It meant that some social reform would be unpopular with these groups and that it had to have strong policies and leadership

In 1893, Kidd asked his boss at the Inland Revenue, Alfred Milner, to help find a publisher for a book he was writing about these arguments. *Social Evolution* was published the following year and sold thousands of copies throughout the world, especially in America where it helped restore religious confidence that had been undermined by explorations in science. This commercial success meant that Kidd could resign from the tax office and spend all of his time writing.

Kidd was invited to Cambridge and met the biologist Roger Fry and his philosopher friend John McTaggart. All three men believed that science and art were both necessary to properly explain living systems. They wanted to find a way to use science to describe their feelings and attitudes regarding the beauty of nature and the structure of human society, but could science alone rid the country of poverty, and would the church stand idly by? Kidd accepted that evolution by natural selection was a force of competitive violence. He understood that it was a condition of any evolutionary pathway, such as our human one, that the less efficient individuals were allowed to fail.

As an unknown self-educated Irishman, Kidd ran into opposition from establishment thinkers in the fields of biology, politics and religion. One of his first adversaries was Thomas Huxley who, in 1893, was still arguing that humans were different to all other species, certainly from Kidd's bees, and therefore were able to challenge many of the normal processes of nature. As though to match Kidd's marketplace style, he used the story of *Jack and the Beanstalk* to restate in simple terms Darwin's idea that opportunism was at the centre of all evolutionary processes. If humans are part of nature then they have to work hard against the rest of nature to survive.

Huxley argued that human society was a garden carved out of nature, but its boundaries had to be defended and its upkeep had to be administered by some kind of authority. This was a paradox. Evolution was the battle in which all combatants (species) would eventually fall. This was one of the first times it was suggested that human kindness – what population geneticists now call 'altruism' – was a feature that evolves. This 'altruism' was not the same as ethics, however. How could there be such a thing as ethics in a natural process such as evolution? Bloomsbury intellectuals enjoyed these discussions about the role of human kindness and some, including Shaw, thought it meant that our

Figure 3.5   Benjamin Kidd (1898)

culture had become an agent of evolutionary change and even saw it as an improvement. However, if the garden were left alone even for a few weeks, it would begin to revert to a very different state of natural equilibrium.

Kidd's critics were from the growing ranks of the new scientific specialists, men proud of their discipline and wary of outsiders. Not surprisingly, the strongest attacks came from the hard science of Galton and Pearson, always fighting for measured analysis and remaining suspicious of anything that was purely qualitative. Karl Pearson had no sympathy with anyone harping back to religion and, in 1895, he published *Reaction!*, a pamphlet attacking Kidd and other leaders of the church for dismissing mathematics and analysis. Not only did Pearson believe in a different kind of evolution, with sudden mutations rather than gradual change, he was also a socialist and wanted a society based on the group rather than the individual. He thought that the scientific evidence weighed against Kidd and that socialism did not have any space for religion. Pearson thought Kidd did not understand because he wasn't a proper scientist but more of a mystic.

Kidd believed in a new kind of ethics, one in which what was right for humans was consistent with evolution. This view appealed to many at the time, especially those who were still outside the socialist movement, men such as Arthur Balfour soon to be Tory prime minister, Alfred Wallace, who especially applauded Kidd's phrase 'equality of opportunity', and Charles Darwin's critic St George Mivart. Soon, after receiving all this attention, his views seemed to vary according to the audience. Similar to his ideas about altruism, he tended to give group behaviour more importance than individualism. It was his belief in an irrational force for evolution that stayed constant. As an idea it was so flexible, however, that it meant different things to different people.

In the final years of Victoria's reign, Butler, Kidd and Pearson began to run out of new ideas about evolution, and even Lankester turned his energy to what he thought was the key to the future: education. He joined the Association for the Promotion of a Teaching University in London, a campaign for the establishment in London of an institution to match Oxford and Cambridge. UCL was short of money, and, in 1887, there was a serious fall in student numbers. Lankester said his chair was 'becoming very rickety' and thought about taking the job of assistant secretary of the British Association. His mentor Thomas Huxley dissuaded him: 'My pet aphorism, "Suffer fools gladly", should be the guide for the Assistant Secretary. You do not suffer fools gladly. On the contrary, you gladly make fools suffer.'[14]

Perhaps it was his tough upbringing that made Huxley a more combative scientist than Lankester. Certainly Huxley had quickly climbed the rungs of the professional ladder and engaged in many difficult battles on the way. He called himself 'a man of science'. He disagreed with the old arguments about evolution made by religious believers such as Butler, Kidd and William Booth of the Salvation Army.

Huxley didn't have much time either for Herbert Spencer who developed his *Universal Law of Evolution* in 1890. This was a theoretical idea that energy is the driver of evolutionary change in nature and in human society, thinking similar to Henri Bergson's contemporary philosophy that was to become popular in France. Spencer's laissez-faire economics and his interest in competitiveness in human groups also became very popular in the United States. They were a set of ideas influenced by Nietzsche and his super-organism, from a competition that paid less attention to the failures than to the successes. Although Spencer's arguments attracted interest from many social scientists, he had more detractors from traditional biology who argued that organisms are not just heaps of molecules, not a set of problems for chemists and physicists alone.

When he was in his seventies, Huxley showed signs of disappointment with his life. He despaired that Darwin had not been vindicated and became dejected about the lack of progress in evolutionary biology. He worried that 'religious passion and rough music' were stopping many people from engaging with Darwin.[15] As though in retaliation, in 1888 Huxley spoke at a Mansion House dinner in the City of London claiming that technical education was an important 'war tax for the purposes of defence'.[16] In the same speech he proclaimed that, 'Industrial competition among the peoples of the world at the present time was warfare which must be carried on by the means of modern warfare, namely scientific knowledge, organization and discipline.'[17] He didn't say whether he was attacking religion or the military or ignorance.

However, Huxley's was a lonely voice. Opposing forces were gathering all around him, both political and religious. William Booth's diatribe, *Darkest England,* came out in 1890; 50,000 copies were sold in the first month. It attacked the elitism of university and technical education and urged members of his Salvation Army, working-class people with no education, to go out and engage in scientific enquiries. For Booth, no understanding or qualifications were necessary to be a scientist. To a crowd of 4,000 people at Exeter Hall in The Strand, Booth argued that his grand scheme would gather 'incontrovertible facts' from the scientific community and make them available to the 'humblest toiler in the great work of social reform'.[18]

Booth's speech encouraged Huxley to join Lankester's campaign to improve London's standards in higher education. With Francis Galton, Huxley campaigned for an Act of Parliament to create a new university. The Albert University, as Huxley wanted it to be called in memory of the Queen's late consort, was to be located near the Geographical Society in South Kensington, and was to become part of Huxley's Normal School of Science and the Royal School of Mines. Lankester was not happy. He accused Huxley of being confused about the place of humans in nature.

Then, in June 1895, there was an unhappy turn of events: Huxley died. Lankester was devastated. The event dampened his fire. He gave a warm eulogy for Huxley, one that came from the heart:

> There has been no man or woman whom I have met on my journey through life, whom I have loved or regarded as I have him, and I feel that the world has shrunk and become a poor thing, now that his splendid spirit and delightful presence are gone from it. Ever since I was a little boy, he has been my ideal and my hero.[19]

4
Insiders and outsiders, 1890s

By the 1890s, Darwin's new biology was attracting interest from people of all backgrounds. There were still the social insiders, such as the Oxbridge graduates from rich families in Kensington, and now there were social outsiders in Bloomsbury with few of the gentlemanly traditions but with strong motivation for studying science. They were looking forward to careers in new fields such as horticulture and pharmacy, as well as in more conventional work teaching or practising medicine. These newcomers to professional life started to make a big impact on Bloomsbury society. This can be seen in a comparison of the lifestyles of two science students who became well-known writers, the insider Roger Fry and the outsider H. G. Wells.

Fry and Wells were both twenty-nine years old when Thomas Huxley died. Despite their very different backgrounds, both men had something in common with Huxley: they had worked hard to bring themselves up in the world and to escape the religious restrictions of their Victorian families. When Fry and Wells graduated with degrees in biology in 1888, they both embarked on careers in the arts where their biological training proved a keen source of inspiration. Though they were never friendly with one another, their stories, their upbringing, training and life choices showed how they both gained from the new social outlook.

Roger Fry's father, Sir Edward Fry, was a leading lawyer in the very middle-class milieu of Hampstead and was convinced that science was the way forward for humanity. In his spare time, he studied mosses and liverworts, work that was original enough to get him elected as a fellow of the Royal Society.

At one stage, Roger Fry thought he might become a professional biologist. On Sunday evenings, it was the custom for Quaker families like the Frys to gather round and listen to stories. Sir Edward would sit at the head of the table and read from works such as *Paradise Lost*. Sometimes the family invited to dinner such men as Ray Lankester, who talked to the family about science. Another visitor was a chemist named Charles Tomlinson, charismatic fellow of the Royal Society, who brought along Chladni's clamp, an invention by which sand formed itself into patterns when a violin was played.

To encourage Roger's burgeoning interest in biology, his father gave him a patch of the family garden just beside Highgate Cemetery. Later, Roger remembered how a stray poppy had caught his attention. Its simple flower was ideal to observe and describe as a biological specimen. It was also very beautiful, its shape and sense of balance set off by the red colour in its petals: 'I conceived that nothing could be more exciting than to see the flower suddenly burst its green case and unfold its immense cup of red.'[1] As a budding scientist, he wanted to look deep inside it and draw, but this developing aesthetic sense, this delight in close observation nurtured by the study of biology and botany, was to take him in a quite different direction: towards the arts.

The other bright young man, H. G. Wells, was also to be befriended by Ray Lankester. Although he had grown up around London and had an interest in the scientific work taking place in Bloomsbury, he was always an outsider, never fully able to pull himself away from his impoverished background. His upbringing was as different from that of Roger Fry as could be. His childhood was dominated by poverty, squalor and unhappiness, and his parents kept a drapery shop in Bromley, on the south-east edge of London. They lived above the shop with their three sons. Bertie, as he was usually known, was the youngest and the most affected by the cramped space and stinking open drains in the back yard. He wrote of his mother, Sarah, that 'in her heart, something went wrong when my sister died ... before I was born.'[2] That kind of tragedy was commonplace in the 1860s and, in good Victorian tradition, was never discussed. Their father played cricket for Kent and claimed fame when he bowled out four Sussex batsmen in successive balls. When he was not away playing cricket, he worked in the shop and drank in the local public house.

Bertie was not impressed by any of this. He spent his time in the public library, reading books about nature and science. Sarah Wells considered the free local school too rough for Bertie and favoured a small private school that cost a penny a day. To save money, he was not sent to school until he was eight; but the cash set aside for his schooling ran out and five years later in 1880, he had to leave. Sarah Wells did not get on with her husband either, and she finally escaped the dismal life in Bromley by becoming housekeeper to Sir Harry Featherstonhaugh FRS at Uppark in Sussex. It was there, in a cupboard, that the inquisitive Bertie found a brass telescope, which he always remembered: the sign of his first scientific interest.

That gentrified country-house style of living appealed to Bertie's mother. It allowed her to practise her strong religious faith by going to church on Sundays, and there she made friends with local people who shared her admiration for Queen Victoria. She arranged for Bertie to serve apprenticeships in another draper's and at a pharmacist, both of which he hated. Despite this, he did have one piece of luck. The pharmacist asked the headmaster of the local grammar school, Mr Byatt, to teach Bertie Latin. Mr Byatt soon discovered that Bertie was an eager and intelligent pupil, much more so than any he had ever taught. This led to Bertie's appointment, three years later, as Mr Byatt's teaching assistant. Mr Byatt encouraged and tutored him to take the entrance exam for the Normal School of Science, T. H. Huxley's college in London. Bertie passed the exam and was soon going to London to study a degree course in zoology and geology.

Roger Fry also had an unusual schooling. He was sent to a new kind of school, one that was looking forward to the new century rather than backward to some proud tradition or high specialism. This was Clifton College, where he was influenced as much by his fellow pupils as by the teachers. His best friend was the timid John McTaggart who was to become a specialist in Hegel and the philosophy of time. Clifton College had several charismatic teachers of science who made a big impact on many of the boys. In Fry's time there, two of Clifton's staff were fellows of the Royal Society and others went on to university chairs in science.

Perhaps inspired by McTaggart or some other boy at Clifton College, Fry became something of a rebel, and so he developed a cautious attitude to the scientists of the day, especially because many of them had been introduced to him in person by his parents. Without the pressure of tradition, most of the students at the college were not expected to go to Oxford or Cambridge. Instead, they were developing different methods and approaches to scientific questions and were showing different political attitudes. Theirs was an exciting moment in history that might lead to new, more liberal ways of living. There was no guarantee, however, that their life would be more comfortable.

In 1885, Fry and Wells found themselves away at university studying science: Fry at Trinity College Cambridge, Wells at the Normal School of Science. For Fry, the place at university had been expected, for Wells it was an amazing reward after much struggle. Both men entered their new life with serious doubts. They were apprehensive about where their studies might lead and regretted that they had not thought more carefully about their choice of subjects. Fry began to think that he really wanted to study fine art and not science as his father had demanded. Wells, for his part, was bitter about the fact that, to attend his institution, he was obliged to join the Church of England: 'The result was that I committed the first humiliating act of my life. I ate doubt ... and lost my personal honour.'[3]

Nevertheless, both men began their university work excited by the new personal, social and scientific opportunities the new century offered. For Wells, classes at the Normal School of Science included systematic zoology, physiology and anatomy. There was also crystallography, which he particularly liked, and stratigraphy, which he did not. He complained that most of the lectures were badly delivered: full of facts and with very little interpretation or even comparison. And he hated exams.

Similar to UCL and King's College, the Normal School of Science and Royal School of Mines offered University of London degrees, but its courses focused on more technical subjects overseen by the government's 'Science and Art Department'. The school occupied the site of the 1851 Great Exhibition in South Kensington and in Wells's day was developing a reputation for providing schools with science teachers. This fitted neatly into the class structure of English society so that the newly educated working class went to South Kensington and the middle classes still went to Oxford or Cambridge.

Wells moved to Bloomsbury in 1884, and lived in a small room in his Aunt Mary's boarding house at 181 Euston Road, now the site of the Wellcome Institute. Part of the attraction of the Normal School for Wells was the professor Thomas Huxley, but unfortunately Huxley left at the end of that first year. Wells complained that the closest he got to Huxley was to open a door for him and bid 'Good day, sir.' The new student worked hard and well, was inspired by all the excitement surrounding evolution, the challenge it posed to so many old conventions, and particularly enjoyed the intellectual discussions at the college debating society. With the friends he made at the Normal School, Wells exchanged books, visited museums and attended the opera, especially Gilbert and Sullivan. Indeed, after a few months, he began to find it hard to concentrate on his work.

Another distraction was his Aunt Mary's daughter, Isabel, an attractive young woman the same age as Wells, and they walked to work together each morning. In the evenings, they wrote their essays at the same table. Isabel was enrolled at the Mechanics' Institute, a small college in Fetter Lane for part-time students and later called Birkbeck College. She enjoyed sharing her educational experiences with Wells and invited him for weekend walks in Regent's Park. For these walks, Wells dressed in his top hat and tails, and his shirt had a faded rubber collar that he tried to scrub clean with a toothbrush after each outing.

At Cambridge, Fry was even more easily diverted from his scientific business and began to admit to his friends that his interest in biology was restricted to the beauty and behaviour of the specimens. Fry joined a group that met at John McTaggart's rooms in Trinity College on Thursday evenings to discuss such undergraduate topics as religion and the meaning of life. The group liked to call McTaggart an atheist who believed in God:

> McTaggart's seen through God
> And put him on the shelf;
> Isn't it rather odd
> He doesn't see through himself?[4]

Fry made such an impression on this group, and on college society more generally, that in 1887 he was elected to a very secret society known as the Cambridge Apostles. Established in the 1820s, the Apostles invited some young men of the Cambridge intelligentsia to join its elite ranks. The family names of the society members were a roll call of the intellectual establishment. They were the young men soon to be seen at the Stephen sisters' Bloomsbury parties: John McTaggart, Godfrey Hardy, George Moore, Roger Fry, Maynard Keynes, Clive Bell, Leonard Woolf, Bertrand Russell, Alfred Whitehead and Lytton Strachey. The first thing Fry did upon hearing that he had been elected to the secret society was to write home to tell his mother about the honour.

Mathematics and biology were popular topics of conversation at the Saturday meetings of the Apostles. Regular attendees included Moore, Fry, Hardy, Russell and Whitehead. This group had endless debates about truth and honesty, and the search for a meaning to life that avoided utopias and the myths of heaven. 'Bertie [Russell] thinks I am muddle-headed', Whitehead would say, 'but then I think he is simple-minded.'[5]

In 1888, Fry graduated in the life sciences with a first. At that time, he was the only scientist who attended Apostles meetings and told friends that he wanted to stay on at Cambridge to decide between science and art. However, he had already made up his mind: all the discussions he had

had at the Apostles meetings made him less inclined to seek out the precision of objective truth. Instead, he was more interested in responding to the needs and feelings of individual people. The powerful image from his childhood of the poppies in his garden continued to encourage him to understand form in an aesthetic, as well as a scientific, sense. He began to believe that life was more than the microscopic and atomic detail of biology; it was also the image and the form, and it changed through time. What mattered was the impact these forms had on the observer. By 1894, Fry was giving Cambridge University extension lectures in art, and writing art criticism for the *Burlington Magazine*.

In 1895, Fry's Apostle friend Bertrand Russell put forward another philosopher for election to the Apostles, G. E. Moore. Moore was twenty-one, looked like a schoolboy and was always serious and reticent in manner. He had a love of clear, sparing prose and common sense. He despised the jargon and difficult language of the German philosophers who dominated the philosophical field and were so hard for ordinary people to understand. Maynard Keynes later said that Moore could speak of love, beauty and truth as though they were as solid and useful as furniture. Moore did not like small talk, or fancy. His personality and prose were both plain, and he admitted to thinking a lot about the basic concepts of

Figure 4.1   Roger Fry, self-portrait (1928)

ethics. For example, while Russell had argued that goodness was what the utilitarians of the earlier generation took to be the pursuit of happiness, Moore argued instead that goodness was not natural to the human species, but that pleasure was.

Fry and Russell saw to it that evolution and the environment were frequent topics for discussion at Apostles meetings. They argued that biology did not have its own set of laws like chemistry and physics but instead was made up of parts from each, together with things from nature's beauty. Moore found this vision of biology very difficult to accept. At one meeting of the Apostles, he shared his worry, observing that Virginia creeper could only survive when there was something for it to climb up. Take away the wall of the house and the plant could not adapt, let alone live: it died. Moore's *Principia Ethica* argued that all evolutionary ethics are flawed, that advanced organisms are not necessarily more advanced morally.

The Apostles regularly pursued biologically inflected philosophical questions: the nature of human intercourse, the enjoyment of beauty in objects and how to find causes and effects of change within complex systems. Few topics were beyond their scrutiny; very little was taboo. At one of the debates in 1899, Moore asked, 'Is self-abuse bad as an end?' He had crossed out an earlier version that ended 'or only as a means'.[6] Moore and the mathematician G. H. Hardy agreed that masturbation was bad, but most did not understand the innuendo in the wording of the proposition let alone the prevalence of the different kinds of sexual behaviour that could be involved. It must have been a heady time for many of them to be able to explore previously taboo sexual questions and to experiment with sexual intimacies between men, intimacies that had been outlawed in Victorian society.

It was also frustratingly difficult to find absolute answers to the philosophical, ethical and social questions that vexed the Apostles. That was why Russell liked mathematics, 'because it is not human'.[7] Yet he also understood that the value of art resided also in its mystery. Moore did not like to use vague words such as 'mystery' and so he disagreed with Russell. As though to emphasise their differences, Russell once asked Moore, 'You don't like me, do you, Moore?' There was a pause that lasted several minutes followed by a simple answer. 'No!' Then they carried on with their conversation at the same pace as before.[8]

Fry viewed the Apostles as his own family. For him, they offered a space in which he would share intellectual expectations for the new century, especially the expectations of honesty, goodness and pleasure. They were an unusual group, protected from the demands of life outside their colleges, but their lack of worldly experience and sense of superiority and entitlement often turned their debates into immature ramblings. It was easy for the group to become immersed in its own self-importance: the members were too insular to make serious advances in intellectual thought. They were not really embracing any of the significant biological advances that were taking place in London and elsewhere.

By 1886, in his second year at university, Bertie Wells was becoming stimulated by the broader social and political implications of the human sciences he was studying. He read widely at first, to gather facts for his zoology essays, but he began to move to the other library shelves in search of ideas he could use at political debates. He believed passionately in democratic socialism and used his knowledge of behaviour patterns from biology to argue that the individual should be merged within the population: that is, the state. He always applied his science to some kind of practical end, usually political. He believed that knowledge was there to improve the human condition, and he wanted to bring these heady new ideas and ways of seeing the world to as wide a public as possible.

Wells, perhaps conscious of the disappearance of the mystical in the post-industrial age, started reading William Blake and Thomas Carlyle. As Carlyle said, 'Men have lost their belief in the invisible. They hope and work only in the visible.'[9] Even at this early stage in the development of the life sciences, most people, including Wells, did not believe that the universe could be reduced to a collection of particles and facts. Instead, they were taught that the humanities were essential to human understanding. Lankester also had an influence on the adventurous Wells. All the students on Huxley's biology courses read Lankester's work, and his book *Degeneration: A Chapter in Darwinism* was required reading. The students also knew Lankester, who still helped with the teaching and spoke enthusiastically about living in those exciting modernist times. Reading Lankester was one of the most interesting parts of Wells's studies in South Kensington. Wells was especially fascinated with an argument that degeneration threatened the middle class because it showed how evolution might bring about a more primitive race. If the politicians didn't watch out, Lankester warned, degeneration would lead to barbarism and racial war. Throughout the 1890s, the question of social progress and the danger of regression were popular topics of conversation across London.

Student politics and reading novels distracted from Wells's scientific studies, and, in the second year, he failed the geology exams. This made his scholarship untenable and forced him to leave the college without a degree. It was a big blow to his confidence and to his ambition to become a professional biologist. He left his Aunt Mary's house in Bloomsbury and moved to practise school teaching in Wales. There, he struggled with his health, persisted with reading the classics of English literature and a year or so later was hired as a tutor with the University Correspondence College to work back in London. At the same time, he studied for his final degree in the evenings and eventually graduated in zoology in 1891. Events in his life then started to move quickly and in unexpected ways.

With the wide-ranging experiences and knowledge that he had gained at South Kensington, Wells wrote two essays in 1890 that reveal his sensitive perception of how art and science influenced one another. *The Universe Rigid* explained how objective science was to give a clear structure to a new view of the world and beyond. Physicists and chemists liked to assume that the world was a closed system, already programmed to go in a preordained direction. It was a view that horrified Wells, and he explained why in this first essay, which turned into an explanation of his atheism. *The Rediscovery of the Unique* had a more subjective emphasis. It looked at individual consciousness, that is, how different individuals feel about being alive. The two essays formed the basis for much of the author's subsequent output, especially the novels that he wrote in the 1890s.

Back to London, Wells had been reunited with Aunt Mary's daughter, Isabel, and they married at the end of 1891. The marriage also marked the beginning of his many affairs. Wells, like a few others of his generation, disliked the compulsory monogamy of the institution of marriage and was determined to find ways of entering into human relationships that did not include possession. Of course, his sexual and romantic experiments on the whole were designed to suit himself and usually caused damage to his partners.

Not long after his marriage, Wells became attracted to one of his students on the university correspondence course, Amy Caroline Robbins, whom he called Jane. Jane helped Wells write his first publication, *Textbook of Biology*, which came out in 1893. The couple took rented rooms near Mornington Crescent in January 1894. They paid the landlady, a Mrs Reinach, a guinea a week for meals and the exclusive use of the two rooms on the ground floor. Two years later, Wells left Isabel for Jane, and this strange relationship survived until her death in 1931.

Wells's *Experiment in Autobiography* speaks happily of these times. There is no doubt that he and Jane were very much in love. They were both busy writing, bouncing ideas off one another. Jane had her biology essays, and Bertie was working on *The Time Machine*, his first full novel. Their two rooms were cluttered and busy with this industry; they had a big box for clothes and a shelf for books and papers. After a few months, Jane found quieter lodgings around the corner at 12 Mornington Road, but unfortunately the new flat had the same claustrophobic atmosphere. Their solution was to escape to work in the British Museum reading room, only a few minutes away, and for walks in Regent's Park.

For some time, H. G. Wells had been finding ways of explaining complex scientific examples in the classroom using metaphors, illustrations and stories, and he realised he had a talent of writing about science for the general public. *The Time Machine*, published in 1894, was an optimistic utopia about a degenerate society of the future in which humans were replaced by a newly evolved fictional species of troglodytes, adapted to underground life. The time taken for this evolutionary process increased from 12,203 years in the first 1894 edition to 802,701 in the 1905 version. Wells realised that the changes needed more time than he first thought. Troglodytes were formed by degeneration, the process that Lankester had taught. The story told that in the far future there was a much-reduced range of biodiversity: a good deal of green slime and a few defensive 'crab-like creatures' were all of nature's beauty that remained. In the year 30,000,000, he found lichens, blood-red sea and a creature with tentacles hopping around fitfully under a dying sun. Wells's

Figure 4.2   H. G. Wells

hope was that this image of extreme degeneration would alert humans to the new knowledge of biology and save them from such desolation.

It was around this time that Wells was introduced to Lewis Hind, a journalist with a best-selling weekly magazine, the *Pall Mall Gazette*. Wells recognised Hind's unusual experience of being placed between science and fantasy, aware that they were living through a revolution in technological advances. Electric lights, telephones and typewriters were changing the workplace; motor cars and steamships were changing transport, and here was Hind, who could write stories and have visions of where it was all leading. The two men became good friends.

Late in the spring of 1894 Wells called to the offices of the *Pall Mall Gazette* in Charing Cross Road to show Hind his latest ideas. He was asked to wait in a magnificently furnished room surrounded by pictures of the *Gazette*'s well-known contributors, W. B. Yeats, Thomas Hardy and Rudyard Kipling. After a few moments Wells heard a painful sobbing from behind a sofa, gave a polite little cough to show his presence and saw the editor of the magazine, the great Harry Cust, embarrassingly get to his feet and apologise to the visitor. Cust wiped away the tears, muttered that he had just had an argument with his lover, and then introduced himself. He said he enjoyed Wells's work, called for his colleague Hind to join them and offered to pay five guineas a time for short stories with a scientific theme. So began a series of thirty-five science-fiction contributions to the *Gazette*. Wells, the working-class boy, had made good. His contributions to the *Gazette* led to introductions to literary agents, and, before he knew where he was, he had joined G. B. Shaw as a drama critic for the prestigious *Saturday Review*, without ever having seen a West End production. Wells and Shaw were to have a lively relationship. Although Shaw had a very different explanation of biological evolution than natural selection, he did go along with Wells's enthusiasm for socialist politics.

Wells and Jane lived in the Mornington Road rooms for another year. Jane typed out Wells's scribbled first drafts and devoted her routine to promoting her lover's self-interests. She lost interest in her own work and abandoned her university correspondence course as quickly as Wells had stopped teaching it. One of their closest friends was Arthur Morley Davies, a palaeontologist whom Wells had met while studying geology, and they often met over dinner. They had similar working histories and they talked about how best to communicate the excitement they felt for progress in science. Davies' *Introduction to Palaeontology* was published twenty years later and became a classic that was still in use in the 1960s.

Within just three years, Wells had written a formidable list of science-fiction stories, including *The Island of Dr Moreau, The Invisible Man, War of the Worlds* and *Love and Mr Lewisham*. They sold by the million worldwide, and by the time he was thirty, Wells was established as the world's leading exponent of science fiction. He was a visionary of the forthcoming century. He foretold not only the influence of the new sciences on the arts in the twentieth century but also the impact on society of new attitudes such as socialism and eugenics.

Wells was one of a generation of artists and writers who considered science and art as a whole and used both to speculate about the future direction of society. Many books by Wells and others attempted to predict a future, offering both optimistic utopias and pessimistic nightmares. Five years later, in July 1900, Wells, now the author of more successful books and firmly established in the literary world of London, received a letter of congratulations about his latest work from the examiner who failed him thirteen years before, Ray Lankester.

> In your last book you tell of the search for the ciliated funnel of the earthworm's nephridium – a little game for which I am responsible. I should so much like to talk with you ... and then let us go and have a chop together in the Majolica Restaurant of your old haunt.[10]

Wells was in awe of his old examiner, and there was no doubt that Lankester still had a strong influence on him. He wrote to Lankester, '[T]o find you among my readers is something of a shock. I really am exceedingly proud of your approval of my work; I only wish I had earned it more thoroughly.'[11] Wells and Lankester met soon afterwards and, despite their different backgrounds and Lankester's status as Wells's old tutor, they found enough in common to become good friends.

Through the early 1890s, Roger Fry was growing frustrated with the rigidity and discipline of scientific life. He was too much of a socialite to follow such a regime. He also believed that in art, as in science, the next discovery was waiting for the next investigator whoever that may be, so he enjoyed looking at the poppies and wallowing in their crimson petals. For him, writing and art were more interesting and challenging than looking down the microscope for clues about how plants work. Like other Apostles, Fry thought it was impossible to find chemical and physical answers to emotional questions about beauty and aesthetics. Therefore, instead of staying on at the Botany School in Cambridge as his parents wanted, he went on a tour of Europe. In Venice he met John Addington Symonds, a well-known gay-rights campaigner. Fry wrote about their time together: 'Symonds is the most pornographic person I ever saw but not in the least nasty ... he has become most confidential to me over certain passages in his life. He is a curious creature – very dogmatic and overbearing in discussion, but with broad views of life.'[12] The spirit of freedom nurtured by the Apostles raised questions, especially about how to deal with the strong opposing force of parental control. Fry and his Apostle friends were challenging generations of family ritual and Victorian convention.

However, Fry still could not decide what to do with his own life. On his return from Italy, a chance meeting with Shaw on top of a horse-drawn omnibus made him think positively about moving away from the past. '[Shaw] took occasion to explain to me what a colossal farce British justice was. Up to then my respect for my father had led me to take his word for it as anything so pure as British justice had never been known on earth.'[13] Fry was determined to break away and do what he wanted: 'I may be a bloody fool but am at least as obstinate as a pig.'[14] He was fascinated by the question of perception and the fact that our senses only related to the world in fragmentary form, never as a whole. He was also interested in the mystery of why nature appeared to be so beautiful to humans. Was this thanks to art rather than the living systems themselves? His debates with the Apostles during the 1890s demonstrated to him the need for both science and art in the appreciation of nature.

It was with these experiences that, by the end of the century, Fry saw a role for himself as one of the first professional artists to have trained as a scientist. To the question of why humans are here, science offered alternative explanations to those given by religion. It challenged philosophers to work out new ethical frameworks to guide us through the technology of modern living. Fry was one of the few people interested in such social dynamics, and he encouraged others to present their views. He met regularly with his friends to discuss the relationship between art and politics, science and society. At one of these meetings, Wells outlined his desire to edit a collection of essays on the subject. For most contributors it was a chance to present their manifesto of the kind of new society they expected, what Wells called 'normal social life'. As we will see in Chapter 8, Virginia Woolf detected that human character changed very suddenly. Maybe this was something similar but more individual and equally catastrophic in the progress of social history.

In his 1905 book, *Modern Utopia*, Wells brought science, art and politics together. Along with many others at the end of the Victorian age, he had a strong belief that science would banish poverty and disease, given the right political support. He had in mind a Platonic world government and fantasised about walking south, across the Alps into order, rationality, beauty, convenience, peace and good health. Together, Wells and Fry saw science as an art, occurring together naturally, like the product of a mathematical equation.

Wells's and Fry's creative use of biology was determined by the peculiar political circumstances of their time. They had the foresight and honesty to realise they were not going to be great scientists themselves, so they chose alternative careers. These new paths engaged their deep sense of enquiry and allowed them to express their frustrations with orthodox beliefs and practices. They were to spend the rest of their lives pursuing their very different aims for society.

5
A new breed of professional, 1890–1904

Ray Lankester wanted passionately to find ways of persuading young people with talent to choose scientific careers. He believed that appointments had to be made on merit and that it was necessary to improve educational methods as well as to increase access for able men and women of all backgrounds. Already, in Bloomsbury, there were several groups of young professionals looking for new ways forward in the biological sciences. They were hoping to build on Charles Darwin's legacy of evolutionary theory and to improve the health and well-being of the human species. Few of them realised the extent of the political implications that were involved with these apparently straightforward ambitions. They were taking on one of the biggest challenges in human history, touching issues of theology, demography, sexuality and poverty.

To have a more widespread impact, Lankester knew that he had to win over the most powerful intellects in the land, minds of men who were mostly at Oxford or Cambridge rather than at the relatively recently established university colleges in London. To prove the point he noticed that throughout his tenure at UCL the brightest of his students stayed for one or two years and then moved to study biology for three further years at Cambridge. Raphael Weldon had done this, starting at UCL in 1876, moving to King's College in the Strand in 1877 and then to Cambridge in 1878 where he got a first in natural sciences three years later.

Another trend that showed up for the majority of Bloomsbury biologists in those years was the enthusiasm of parents. For a start, it was parents who paid the fees. That was a significant sum, about the same amount of money that an unskilled labourer earnt in a year. However, because biology was a new subject there was no clear profession at the end of the college training, except of course in medicine, where most of Lankester's students ended up. To see an education in biology through to a job in that field demanded continuing academic work and psychological encouragement as well as the risk of yet more need for financial support. In so many cases it was the student's parents who provided most of this help. Weldon's parents were journalists, and his friend Pearson's father a lawyer, pleased to see his son work as a mathematician.

To spread these ideas, Lankester applied to become professor of zoology at Oxford, where he thought he could influence another group of professors about the urgency for educational reform and the importance of evolutionary biology in the training of modern scientists. He moved to Oxford to take up the post in 1891, to an old-fashioned biology department with no facilities to develop laboratory skills. These had to be started from scratch, alongside new courses about the microscopic structures inside cells and the environment outside organisms. Also, there had to be more biological understanding of psychology and a deeper explanation of the growing interests in human race and breeding.

Lankester was worried about the political consequences of the emerging eugenics movement. From their very different perspectives, Dr Bryan Donkin and H. G. Wells also thought that it raised important issues, and they agreed that Lankester was a good person to point out such dangers to the public. He tried to integrate as many scientific ideas as possible into his projects, even when it meant that he had to deal with people he didn't like. One problem was his bad temper that meant sometimes he found it hard to maintain good relations with people. He did his best to keep personalities and personal ambitions out of the fray, and increasingly he took the middle road politically, but he vigorously rejected extreme ideas such as state control and eugenics. Unlike Fry, who wanted to stay with the establishment, and unlike Wells who wanted to support groups of victims, Lankester was unhappy with the thought of going too far in either direction.

It did not take long for Lankester to confirm what he already suspected: that collegiate Oxford was a closed and heavily defended society. It had no tradition of teaching biology, and the methods used were to cram uninterested students for the examinations. 'I cannot fall in with it', Lankester complained to his brothers when he visited them for Christmas.[1] Eventually he lowered his expectations and focused his work on what he knew he was good at: renewing the exhibits at Oxford's Museum of Natural History, starting laboratory-based dissections and new courses in biology for medical students. Aware of the importance of social mobility, he campaigned for open scholarships to students in all subjects.

Lankester was soon well known at Oxford for this hostility to the many unmotivated students. He was also unpopular with his many complacent colleagues who had spent their whole adult lives at the university and who took their positions for granted. They didn't care that the facilities for science were so bad and were bemused when the new professor decided to reorganise from scratch everything in his care.

However, just as he started to bring about change in the zoology department, other things started to go wrong. He was unhappy with his own lifestyle in Oxford. He missed his old colleagues and students in the capital and was anxious about making a home out in the provinces. The time and effort that these educational ambitions demanded meant that he had little time left for his research in marine biology, and his standards slipped. An article he published in *Proceedings of the Royal Society* describing a new species of freshwater jellyfish was challenged, and the editors had to publish an apology on his behalf because someone else had got there first. It was the kind of slip that Lankester could ill afford to make. His many enemies were going to be quick to pounce with criticism. One of these was Samuel Butler who decided to criticise the jellyfish work and tried to use it as evidence against natural selection. In addition, Lankester was publicly lampooned by one of his few Oxford friends, Worthington George Smith, in a cartoon, depicting him as a jellyfish and entitled: 'Scientific Unworthies No. 1: A Dirty-Water Medusa'.

Then came yet more bad fortune. In 1894, Lankester became engaged to a Mary Corbett, about whom very little is recorded. Just before the wedding ceremony there was a quarrel, and she broke it all off. Lankester was devastated. He feared that he would never get married. For a year or more he poured out his emotions into a private journal.

A further disaster came the following year, in October 1895. The story goes that after one of his regular Saturday evenings spent at his club in Piccadilly, Lankester walked out into the street and saw the police drag a screaming woman out of a cab. Lankester went to ask what was going on and more police soon arrived. They asked Lankester to move on, and he refused, only to be quickly arrested himself and charged with obstruction. To make matters worse, he hired one of the most high-profile advocates of the time who then chose to argue with the magistrate. His client was fined £10 and bound over not to break the peace. Not content to let the already over-publicised matter rest, Lankester wrote to *The Times* and asked Lord Salisbury, the home secretary, to intervene. Salisbury refused.

***

Meanwhile, during Lankester's Oxford years, three young botanists had started work in Bloomsbury and went on to make important contributions to biology. They were Frances Oliver, Arthur Tansley and Marie Stopes. Oliver came from the strongest of all biological environments, at Kew, where his father was keeper of the Herbarium, and at UCL, professor of botany from 1861 to 1888, when his 26-year-old son took over the professorship. Frances was an admirer of Leslie Stephen through their interest in mountaineering and they had both studied at Cambridge, thirty years apart.

Oliver loved to find good examples of evolutionary lineages for his teaching. He had been impressed by his students' attentiveness whenever he mentioned *Archaeopterix*, the 150-million-year-old fossil that seemed to link the dinosaurs with modern birds. The first such fossil had been found in Germany in 1861, and more specimens from rocks of different ages in the same region were beginning to tell the story of how evolutionary pathways developed. Equally, Oliver was excited when he and his colleague D. H. Scott found another example of ancient relationships between what were thought to be different species of plants. They found the same distinct glandular cells on two very different kinds of fossil plants: a fern frond and a simple seed. Did

Figure 5.1 Leslie Stephen (1902)

the same kind of glands mean they were from the same plant? At first Oliver thought that this would be impossible because modern ferns do not have seeds. Later, however, he found that they were connected. This proved that the plant belonged to an extinct group that grew in the humid coal swamps of the Carboniferous Age. Scott and Oliver called these plants Pteridosperms, giant tropical ferns that reproduced by a simple attraction of pollen to the exposed ovules. It was a revelation and a breakthrough to discover whole groups of plants as well as animals that were extinct.

Oliver was an environmental botanist interested in how plants could grow in unstable places like sand dunes. He was one of the first to plant the hybrid *Spartina townsendii* for the purposes of land reclamation. Oliver knew it was going to be decades before the main processes of plant growth could be understood because plant communities took that scale of time to become established especially on new shingle, and many of the processes involved were obscure. Moreover, it took decades to monitor ecology, which meant that this kind of work was overshadowed by quicker laboratory experiments on other problems, ones that yielded data for the fashionable techniques of statistics and biometry. The new study of ecology was not going to give any fast solutions.

Frances Oliver gave public talks with Lankester about their concern with the unhealthy conditions in many London streets and together focused some of their research on the effect of atmospheric coal dust on vegetation and on water-borne diseases. Thick fogs were becoming common during the winter, and drinking water was still not reliably clean. The work encouraged Oliver's contemporary Arthur Tansley to look at the effects of salt spray on the plants that grow on sand dunes, shingle beaches and in salt marshes. Tansley looked at the interactions between the plants and these extreme environments to see how certain plant formations built up on the coast. This work needed constant monitoring over many years. Tansley, as well as Oliver and groups of their students, made seasonal measurements on the north coast of Norfolk where the shingle beach system was continually being destroyed by storms and later redeveloped. They measured changes in geography, meteorology and the flora and fauna. In their experiments, some of the earliest to be carried out in the field of ecology, they were hoping to find patterns in plant and animal migration, and maybe even evidence of adaptation.

Tansley and Stopes both came from the aspiring lower and middle ranks of the Victorian social spectrum, the former from a local household of skilled working-class parents and the latter from a family of brewers with an eccentric interest in scientific archaeology. Lankester was thrilled at their success and used this to make sure that more people of their background were being appointed.

Arthur Tansley was born in 1871 at a house in Regent's Square in Bloomsbury. His family shared it with several others, part of the Duke of Bedford's estate, notorious for charging high rents for squalid cramped accommodation. For example, 1,700 poor people lived in two blocks in nearby Little Coram Street. Despite these conditions, Tansley was determined to 'come good' and had plenty of encouragement from his hardworking parents. Across the same square there was a different optimism in Edward Irving's infamous Scottish church where members of the congregation could be heard speaking in tongues, preparing for the end of the world.

Tansley's father was a skilled carpenter who took science and learning very seriously. Working with wood meant that Tansley Senior knew a lot about the structure and properties of timber, and he still lectured on tree anatomy at the local working men's college in Great Ormond Street. The young Arthur himself became a student at the college and learnt to emulate his father's enjoyment of hard work and love of nature. Whereas the children of grand liberal families of the earlier middle class had sought openness and individualism, Tansley was more cautious and measured. He was looking for another kind of salvation, and, though he was trying to understand nature as a whole, he tended to answer one question with yet more questions. He became adept at this and began to ask himself why he was becoming so troubled with self-doubt. To try to understand, he attempted to read the few books devoted to psychology and found nothing to help. At the end of the nineteenth century, psychology was a very young subject.

As a young lecturer in the botany department at UCL, Tansley became a popular figure within the students' union. He was union vice-president in 1898 and often gave talks on general science, such as 'The Origin of Death'. He began with the simple observation that an amoeba splits into two new animals at the end of its life, thus never dying, but his main theme was the importance of environmental change in driving evolution. It explained how and why each species existed; it was the key to how unique sets of physical factors caused each adaptation to change. Arthur Tansley's reputation and confidence grew, and he helped Herbert Spencer to revise his work on plant morphology and physiology for the 1899 edition of *The Principles of Biology*.

It was becoming clear to Tansley that the cultural importance of evolution spread beyond science into religion, the arts and politics. This shift meant that biology was moving further away from the romantic descriptions known as natural history that were common in the mid-Victorian period. At one pole quantitative experiments were given priority while at the other the politicians and theologians wanted data they could interpret to favour their own ends.

For most people in Europe, what mattered was that their lifestyle changes were material, enabled by an increase in quantitative data. Only a few decades before, Francis Galton bemoaned the lack of such data, making his early attempts to develop statistical interpretations very difficult. Suddenly, just before the new twentieth century, measurements were coming from many kinds of new technology, and they encouraged other changes. Writers were expressing their feelings about themselves, artists were reconsidering colour and the meanings of an image, and biologists were beginning to analyse measurements: modernism was having a wide impact.

Tansley lectured about these influences on the environment to a new charity for social welfare established at Commercial Street in 1884, showing how plants adapted to changing environments. This was what Lankester's friend Ernst Haeckel was calling 'oecology', a word Tansley was to anglicise later to 'ecology'. Interested in the design of biological structures, Haeckel, Lankester and Tansley appreciated the beauty of the interactions between species, not only the shape and colour of the structures but also how they changed on different timescales. The three men marvelled at how so much in nature fitted together. They felt humbled by how small an impact humans had made upon that complex whole. They knew from experience that examining the environmental influences on species was something completely different from examining specimens in the laboratory. Something very like this holistic ecology was part of the tradition of the romantic English naturalists like John Ray and Gilbert White, who had looked at nature in the wild and got cold and dirty in the process. For the new ecological scientists, studying the environment was just as important for understanding life as analysing inheritance and inner structure.

Arthur Tansley's work on Oliver's salt marshes at the UCL field centre in Norfolk was breaking new ground, linking plant physiology to growth in these constantly changing environments. Meanwhile, his friend Marie Stopes, who was studying the geological history of similar environments, showed that ancient plants had adapted to environmental changes many times through their history. This was also exciting new work and for the first time showed how ecology played a major part in shaping evolution. After a few years both Tansley and Stopes possessed good examples to show how dynamic environments effected several groups of organisms, both living and extinct.

This kind of biology overlapped with other disciplines such as chemistry, physics and geography. Like many of his contemporaries, Tansley was interested in examining the effects of climate on growth and wanted to incorporate data from as many disciplines as possible. For this reason, he jumped at the chance to take an overseas study tour. He spent the greater part of 1900 and 1901 collecting research material in contrasting environments from Ceylon, the Malay Peninsula and Egypt. He was fascinated by the very different plant and animal communities he saw in these places. In the process of comparing the different specimens and climates, he developed a special interest in the structure and physiology of fern-like plants. His work in this area was to get him elected to the Royal Society.

Thanks to his Asia trip, Tansley also realised how little was known of the flora and fauna of countries outside Europe. He brought together a set of proposals for making standard surveys of the plants growing in different climates and landscapes. Comparing these data with earlier records from the British Isles showed changes from what had been observed earlier. Tansley presented his results to the 1904 meeting of the British Association, and the new science that we now call ecology was officially born. Ecology soon developed applications in agriculture and began with attempts to survey land for agricultural and sustainable purposes, to make species lists alongside soil and weather surveys. Scotland had been extensively surveyed in the 1890s, and Oliver was making large-scale beach surveys in Norfolk, but there was a lot more work to do. Tansley's talk to the British Association led to the establishment of a Committee for the Survey and Study of British Vegetation, with nine active field surveyors led by Tansley himself. Each member of the committee prepared a guidebook to their study areas, which included lists of species, measures of abundance, and information about the soil, the geology and the weather throughout the year.

A few years earlier, in 1901, Tansley had taken £130 of his own cash to a small printing workshop just off the Tottenham Court Road and signed a contract to print a monthly academic journal about ecology. He called it the *New Phytologist*, and the first edition was published in January 1902. It soon gave the new subject, and Tansley's thirst for survey data, respectability and weight within the scientific community. In 1906, he accepted a coveted lectureship at Cambridge and started to plan a five-week survey of the British Isles that comprised his committee's first International Phytogeographical Excursion. Tansley's father would have been proud of these accomplishments for the Tansley family had come a long way from the working men's college in Great Ormond Street.

With the Cambridge lectureship, Tansley was becoming famous for describing different kinds of habitats and their ecology. At the same time, he was infamous for not being able to remember the names of plants when he was out in the field. His ecology was based on the idea of a gradual succession of flora and fauna to a stable climax, but this idea was never placed at the front of scientific advances in evolutionary biology. At that time, no one seemed to care about adaptation to the environment or climate change, let alone catastrophes like tsunamis and meteorites hitting the earth and causing havoc. Instead, ecology was seen as a tool the Empire could use to produce more food and mineral resources. For instance, Marie Stopes' studies of coal helped find the most productive form of energy readily available at the time of the First World War.

As experts in a new age, Lankester, Tansley and Stopes felt it was their duty to encourage public debate about the social issues raised by their scientific work. In 1909, Lankester wrote to H. G. Wells with a preview of a lecture on overpopulation that he was about to give. 'This cannot go on. Man must come to a limit. Then the real science will come in.'[2] He wrote that human numbers were threatening to turn the world into a sort of formicarium or agaricarium, like an old cheese full of mites. Population had to be controlled; atomic motors and land reclamation from the deserts would be insufficient.

Some years later, when Virginia Woolf created a rich portrait of Bloomsbury social circles in her novel *To the Lighthouse*, she named a scientist in her group Charles Tansley. Her protagonist Mrs Ramsay describes Tansley as being far from a polished specimen:

> She was telling lies he could see. She was saying that she did not mean to annoy him, for some reason. She was laughing at him. He was in his old flannel trousers. He had no others. He felt very rough and isolated and lonely. He knew that she was trying to tease him for some reason; she didn't want to go to the Lighthouse with him; she despised him: so did Prue Ramsay; so did they all. But he was not going to be made a fool of by women, so he turned deliberately to his chair and looked out of the window, and said, all in a jerk, very rudely, it would be too rough for her to-morrow. She would be sick.[3]

Yet, this was not how the real Arthur Tansley felt about leaving his class behind. For he could never share his new scientific knowledge with his uneducated parents, and this must have made him feel lonely and isolated from them. This would not have occurred to Woolf because her own father was a leading member of the London literary elite. Her view of

Charles Tansley was that of a poor figure seen by a privileged person. It was what Woolf often saw from her own world, and it caused her to fear that her own elitist existence was entering a phase of decline and possible extinction. Woolf, however, remained fascinated by Arthur Tansley's work. Perhaps her own childhood interests in natural history made her regret that she had not followed it through as Tansley had.

Oliver and Tansley began to realise that important work on ecology was already being done out in the English countryside by groups of amateur enthusiasts, all under the name of natural history. Tansley brought a number of these amateur groups together and had them recognised by the British Vegetation Committee, compiling regional surveys of plant distribution. In 1911, this became the British Ecological Society, and, in the same year, Tansley's *Types of British Vegetation* set an international standard for descriptions of world environmental types. In empires such as Britain's, examining the world's resources and organising their economic importance was an important task. With ten members from Europe and North America, an excursion finally got under way in 1911 to make more surveys. It went through England, Scotland and Ireland, mapping species distribution in order to find migration pathways and other ecological trends in response to changes in climate, geology and human history. This was pioneering work that helped ecology and environmental biogeography develop together, and it was happening at the same time that genetics and biochemistry were advancing knowledge about the inside of animal and plant cells. The study of anatomy and morphology so familiar to Lankester's generation was getting left behind, and much of the new effort was going into ecology. It was no coincidence that these new studies of interactions between organisms and their environment came out of the upwardly mobile Nonconformist London scene.

That energy was from a new kind of scientist: highly motivated, often working-class and sometimes female. They were open-minded about where their new approaches to evolution might lead and ambitious in their approach to new techniques. The future of their scientific research depended on them working together as a group. Such Bloomsbury scientists took gender equality seriously, but it was still difficult for women to gain any opportunities in practical science: Lankester and Oliver were about the only senior scientists at all sympathetic to admitting women into laboratories and onto degree courses.

One of the most academically talented of Tansley's students was Agnes Arber, just eight years younger and sidelined by unimaginative male colleagues who did not like the way she kept raising difficult issues. One of her frequent questions was whether the ecosystems that Tansley was beginning to describe were self-organised. Another was why a particular structure exists in one species and not another. Arber was as much interested in function as structure and wanted to see the whole picture before coming to any conclusions. She had sensitivity, which most of her male colleagues missed: a grasp of the whole, a bird's-eye view of all the living world that she could see. She argued that every individual organism was an essential part of the integrated whole in any living system: this, she proposed, was how the ecosystem functioned.

Arber's broad view and questioning approach fitted some of the new social thinking of those times, but it was met with indifference among the growing number of biologists who measured things. When she suggested that 'urges', 'endeavours' and even 'perseverance' might explain the compulsion that makes living things work, very few scientists took her seriously. Her idea that such inherent urges drove biological development left many options open that had to balance with the harder physical explanations. It was becoming clear just how deeply the biological sciences were being influenced by the hard quantitative sciences of physics and chemistry. What made Arber and Tansley different from most of their contemporaries was their persistent interest in the whole spectrum of factors that might conceivably influence evolution. They were consciously linking the old and new attitudes and were very aware of the mix of interests involved in studies of evolution.

Another in this same group of students in Oliver's botany department was Marie Stopes. A year younger than Arber, Stopes was also taught by Tansley, and they all worked in the same laboratory that had a gallery and a skylight, just above the Slade School of Art. There they prepared thin sections of different tissues for examination under their microscopes, measured physiological tolerances to changing concentrations of dissolved radicals and discussed the importance of new methods of preparing specimens. They drank tea up in the gallery and argued about the colour and geological age of the first flowering plants. Just a week before Stopes' doting father died, at the end of 1902, she received a postcard from her admiring professor at UCL: '1264 gets Bot. 1st Cl. Hons. Geol. 3rd Cl. Hons. F.W.O.' Stopes, now a first-class graduate, rejected the advances of her male admirers and concentrated on the sex lives of plants, particularly early flowers found as fossils.

Stopes soon became known as one of the most adventurous characters in Edwardian Bloomsbury, being the first to break through many sexist barriers. Stopes had had a rigid Scottish upbringing steeped in Protestant convention. Her mother was the first woman to qualify at Edinburgh University and was passionate about women's suffrage and minority

Figure 5.2   Marie Stopes (1904)

causes. Later, Marie would say that she thought that her mother was cold and controlling and gave her husband, Henry Stopes, an unhappy life. Henry was a brewer and an architect, though his architecture business ended in bankruptcy, partly because he put so much of his energy into his daughter and his hobby of archaeology. He spent as much time as he could in the quarries at Swanscombe looking for fossils of early humans.

Not to be outdone by her friend Tansley's new journal about ecology, the *New Phytologist*, Stopes began her own very different *Sportophyte* newsletter with funny poems and short essays.

> Last night as I lay sleeping
> There came a dream so fair
> I stood mid ancient Gymnosperms
> Beside the *Ginkgo* rare.
> I saw the Medulloseae
> With multipartite fronds,
> And watched the sunset rosy
> Through *Calamites* wands.
>
> Oh Cryptogams, Pteridosperms
> And *Sphenophyllum* cones,
> Why did ye ever fossilise
> To Palaeozoic stones?[4]

These were productive times for palaeontologists wanting to describe new species of extinct plants and animals. Marie Stopes did not want for good specimens. She was encouraged to study plant evolution by Oliver and Tansley, and after graduation gave it all of her energy and creative application. Early on, she identified the recently found 100-million-year-old fossil palm leaves from quarries in the Chiltern Hills. She undertook major studies of the older fossil leaves and seeds from the Coal Measures of Lancashire. She examined the anatomical cell structure in order to make accurate identifications and to reconstruct the ancient ecology of which these unfamiliar plants had been part.

During her analyses, Stopes became aware of the biological importance of sex. She thought that by explaining the biology simply, ordinary people would be able to plan the size of their families more easily. This was especially important for young women at the time, who were beginning to fight for political and social recognition. In 1905, Stopes began her crusade for women's rights when she became the first female lecturer in a British university, at Manchester. She used the Eugenics Society to campaign for more understanding of the role that women could play in adjusting human population levels. Like Agnes Arber she sympathised with the Malthusian League. Now she, too, was excited that the new biology might come up with new ways of controlling population and ending mass hunger.

Meanwhile, Tansley had taught himself German and could read the recent publications of Warming's *Plantesamfund* and Schimper's *Pflanzen Geographie auf Physiologischer Grundlage*. These books developed the concept of plant communities and described the relations between plants, soils and climates that Tansley regarded as the foundations for plant ecology. In 1903, he married his former student, Edith Chick, and they remained close friends of Marie Stopes even after they moved up to Cambridge.

Another of Tansley's early London students was Bernard Hart, a medical student who went on to become a well-known psychiatrist. He worked at some of London's most notorious asylums, an experience he drew on in his popular book *The Psychology of Insanity*. Tansley was captivated by Hart's accounts of his work with patients in the consulting room. He enjoyed discussing its biological significance with Hart, the opportunities that Darwin had predicted in the last pages of *On the Origin of Species*: 'In the distant future I see open fields for far more important researches. Psychology will be based on a new foundation, that of the necessary acquirement of each mental power and capacity of gradation. Light will be thrown on the origin of man and his history.'[5] If ever there were a challenge for Tansley to spend more time studying psychology and thinking about its biological importance, this was it. Lankester's ambitions for psychology also sprang from this famous quote, and he encouraged Tansley and Horatio Donkin that the subject had great potential.

By chance, Donkin and Tansley represented the fledgling division that had already appeared in this new discipline of psychology. The medical sciences accepted Darwin's challenge to explore psychology's scope and mechanisms, and by the 1880s it had laid the foundations of neurology and psychiatry as distinct specialisms in the treatment of mental illnesses. For some conditions, such as 'hysteria', distinctly non-medical treatments such as hypnosis were being used even by the most respectable practitioners. None of these clinicians could find any clear scientific explanation for how such methods worked, and their credibility began to suffer in the new age.

One medical practitioner who was uneasy about some of the treatments being used was Donkin himself. He had written a description of hysteria for the 1892 *Dictionary of Psychological Medicine*. He thought it was significant that along with anorexia, hysteria was a common ailment for aristocratic and intelligent young girls, many of whom were expected to be completely devoid of libido. Young hysterics even became celebrities of a sort, 'performing' their attacks under hypnosis, in front of audiences and for medical study. To help the patients, electrical vibrators were marketed from 1883 as a cure-all for women, and as pain relief for men. The advertisements described how much happier, healthier and more vivacious women would be if they bought these massagers for their face and head. Donkin's scepticism about such treatments was taken seriously because it was based on the theory of sexual reproduction in all living systems. For the first time sexual matters were openly discussed, and many people had the confidence and means to live and think in a more informed way.

***

In 1898, Lankester took the job as director of the British Museum (Natural History). He was hoping that, free of the traditions of Oxford University, he would be able to show the public more about how life worked. He wanted to use the new museum in South Kensington, with its colour and diversity, to inspire a new generation of scientists. Instead, however, he was confronted with outdated attitudes, not very different from those he had met at Oxford. He soon found that the museum was run by a committee reluctant to accept the changes brought about by new communications and other new technologies.

Few of the committee members seemed to be very aware of the increasing importance of science, and biology in particular, in British life. The committee wanted to continue its old ways of working, still using the big museum building in Bloomsbury, quietly protecting the collections and proudly waving the flag of empire and nationalism with new expeditions abroad. The rigid attitudes that Lankester encountered in the museum precluded reform. He wanted to make a new catalogue of the collections and was refused help by most of the curators. He wanted to have control over new acquisitions and was overruled by the accountants. He wanted a lecture theatre and was refused by the architects. He was even told by the management committee to change the colour of the walls in one of the corridors. He turned to the chairman and said loudly of the old paint, 'It won't come off.'

Nevertheless, Lankester was always observing, experimenting and collecting data, intentionally following Darwin and Huxley's example with accurate and thoughtful accounts of new species, their form and lifestyles. There were others who wrote about evolution who had never used a microscope and had never been out into the field to explore nature in its real environments: Lankester did not have time for such men and left them to their pontificating in Bloomsbury while he moved on to the new site in South Kensington. He believed that one needed to get one's hands dirty.

Once there, and try as he would, Lankester became even more frustrated with his own failures to find useful clues about evolution in his classical observations of new species, and he lost his temper with others who were equally unsuccessful looking elsewhere. He was still determined, however, to help find the elusive agents of heredity that would support the campaigns of Darwin and Huxley, and there were still plenty of places to seek them out. He was attracting attention with his studies in embryology, but there were not many new developments and increasingly he found it difficult to remain optimistic. Whether he liked it or not, without new evidence Darwinism was moving into decline.

6
A new breed of biologist, 1900–10

During the first decade of the new century, Bloomsbury scientists were at the centre of a number of ground-breaking scientific advances, including the birth of genetics and ecology. This progress gave Ray Lankester a warm sense of achievement, a feeling that his academic mission had been worthwhile, that the improvements to science education really had opened up careers to people from ordinary backgrounds. The wider ambitions of other Edwardian leaders also allowed groups such as the Fabians, the Eugenics Society and the Men and Women's Club to recommend policy changes to the politicians, leading to different approaches to the problems facing those living in the city slums. Across different parts of Britain there was an optimistic atmosphere that scientific advances were steadily improving the quality of life.

Lankester's own student of marine biology, Raphael Weldon, had succeeded him to the chair of zoology at UCL and had become highly regarded as a teacher: 'Seldom is it given to a man to teach as Weldon taught. He lectured almost as one inspired. His extreme earnestness was only equalled by his lucidity. He awoke enthusiasm even in the dullest, and had the divine gift of compelling interest.'[1] Weldon also became well known through his campaigning against the amalgamation of UCL and King's College into a single 'Albert University'. During vacations Weldon worked hard at the Marine Biological Association laboratory at Plymouth, which Lankester had founded in 1888. One of Weldon's projects there lasted several years and involved him collecting thousands of shrimps from their natural habitats. He took several measurements for each one and plotted them to show how characters varied across the species' population. His statistical analysis of these data showed the changes were slow and steady.

Weldon's contemporary in studying biology at Cambridge, William Bateson, didn't altogether believe this conclusion. He had earlier evidence to show that species evolved more quickly than Darwin had assumed and that sudden mutations were caused by sudden environmental stress. Bateson had collected shrimps in the rapidly drying-out Aral Sea and found many new species living in water with very high salinity. This explanation didn't impress Weldon, who reacted by writing, 'The questions raised by the Darwinian hypothesis are purely statistical, and the statistical method is the only one at present obvious by which the hypothesis can be experimentally checked.'[2] The statement was affirmation that Weldon had moved to Bloomsbury where he was using the statistical methods of Francis Galton and Karl Pearson on his own data and interpreting in his own way. Here was a useful difference of opinion. Weldon, and the statisticians, had curves showing that change was slow and smooth, while sceptics of Darwin's gradualism, like Bateson, had evidence of sudden responses to stress. It was what he called a saltation, or now, a mutation.

One of the first sceptics of gradual evolutionary change was Herbert Spencer, and he had arranged an informal meeting to discuss the differences. It was at the Savile Club on 9 December 1893. Together with Galton, Wallace and Lankester, Weldon presented his case for obtaining more evidence by conducting statistical enquiry into the variability of organisms. The meeting agreed that the statistical method was a promising way to check Darwin's hypothesis experimentally. Later, the same group became the Evolution Committee of the Royal Society. The work confirmed that Lankester and Tansley were right to look at the whole organism and the physical environment for evidence. In order to carry out thorough statistical analysis, it was essential to include data from as many parts of the system as possible though it was agreed that the results could never be better than the quality of the incoming data. Most other investigators were set on the new mission to discover the smallest part of the cell, expecting to join with the chemists and physicists in finding mathematically defined laws. It frightened Lankester that both directions of study were so vast. He realised that understanding the mechanism of evolution was going to take a long time.

Lankester wanted to monitor the progress of these different research projects in biology. He took advantage of Spencer's initiative by inviting the same Evolution Committee group to a series of monthly dinners. A sudden fall in student numbers had made him afraid that many bright young people were being seduced away from science and evolutionary biology in particular by poor employment prospects. He wanted the committee's help to find a way through the impasse: just where was biology heading? He also wanted to discuss new ways of studying evolution. Were the elusive hereditary agents going to be found by looking down the microscope, by looking out into the environment or even by formulating a mathematical equation?

Lankester was too deeply entrenched in marine biology to do other research, but he never gave up hope that he might find a new way to prove Darwin's theories. Looking at how different species developed was hard work, and brought little reward. He was often depressed as a result. Offering some hope was the biometric work coming from Galton's laboratory, often helped by two mathematicians who came to the monthly dinners, Alfred Whitehead and Bertrand Russell. Whitehead and Russell were then preparing their monumental *Principia Mathematica*, but they did not find any parameters from Lankester's descriptive biology to fit into their quantitative methods.

Russell was confident that eventually he would understand evolutionary biology objectively. He expected a clear beginning and end that could be joined together using mathematical equations. He saw all knowledge as a single language of science, leading to a single truth, a Law of Life being verifiable by experiments and expressed in numbers. The older naturalists such as Lankester and Alfred Wallace did not accept that life could be described using equations, and they ridiculed the link the mathematicians were making to philosophy.

In Cambridge, William Bateson was determined to find experimental support for inheritance. He had never been impressed by the apparently random lottery of competition at the centre of Darwin's creed, and he could see no evidence either for the units of inheritance being particles (or what we now call genes). Instead, he had a theory that hereditary information was transmitted by sound waves or other vibrations. There was no positive evidence for this, either, and Bateson looked in vain for some kind of experimental verification.

In London, and with equal determination, Francis Galton and Karl Pearson were still analysing the data they had collected over the years about health and family backgrounds. Neither the genetic nor the biometric approach had any new explanations for how life works. All that the two groups were able to agree on was to monitor breeding patterns in experimental animals and plants. Galton, Pearson and Weldon wanted to extend their experiments to humans. Support for this approach came from Russell who was already applying Pearson's statistical reasoning and the principles of Galton's law of ancestral heredity. Russell advocated social action with direct payments from the state to 'desirable' parents, while the poor sections of society, those who reproduced much faster than the wealthy, got nothing.

Lankester's reaction to Russell's scheme is not recorded, though the rift did highlight a growing split between biologists who measured and those who observed. While Russell was all for measurement, Lankester insisted that both were necessary, and for a time the two stayed in their separate corners. Along with Whitehead, Russell advocated three requirements to explain the history of life: a concept of infinity, the flexibility of choice and the desire to reduce explanations to the smallest component. They argued that this hard and optimistic programme raised the stakes for biologists by pushing qualitative description to one side in favour of biometrics.

***

Then, early in the new century, there was an important advance from an unexpected discovery, and it led to the creation of the discipline soon to be called 'genetics'. Early in the morning of 8 May 1900, the forty-year-old William Bateson caught the train from Cambridge to London. On the journey Bateson happened to turn to some papers that he'd bundled into his bag, and there he found an unread reprint of an article published thirty-four years earlier. The article, by a monk called Gregor Mendel, discussed the inherited characteristics of peas over several generations. Bateson found it fascinating. From Liverpool Street Station, his taxi drove through the dirty Bloomsbury streets, and he realised that he had just made an important discovery. Some of the biologists and medics who worked nearby would also have understood the contents of that obscure article had they known of it, but he was going to claim the realisation for himself.

Bateson, then a fellow of St John's College, had been invited to give a lecture to the Royal Horticultural Society in Chelsea about 'Galton's Law'. The lecture he had prepared was based on the conclusions of a recent analysis of the white and yellow patches on Basset hounds that suggested that parents contributed equally to their offspring's inherited matter. However, here was the earlier work by Mendel that showed a different kind of pattern from one generation to the other, a more complex mix than the expected 1:1 ratio. Legend has it that Bateson changed the topic of his lecture to share Mendel's work with his audience at the Chelsea Society. He began with a prediction: 'The Laws of Heredity will probably work more changes in man's outlook on the world, and in his power over nature, than any other advance in natural knowledge that can be foreseen.'[3] The audience remained silent, unaware of the importance of Mendel's pea-breeding experiments, let alone why such dull work should be presented with excitement. It was an uncanny repeat of the silence after the Darwin and Wallace paper that had been read out to the Linnean Society forty-two years earlier. Wanting to share his discovery further, Bateson alerted Galton to the manuscript: 'Mendel's work seems to me one of the most remarkable investigations yet made on heredity, and it is extraordinary that it should have got forgotten.'[4] Like the audience in Chelsea, Galton did not respond, and it took several years for him to appreciate the significance of Mendel's work. To be fair, it was easy to miss the point of Mendel's obscure experiments. For the uninitiated – that is, the vast majority – the changing shapes of peas from one generation to another were hard to associate with the search for Darwin's missing units of inheritance.

With no one showing any interest in his rediscovery, Bateson was left feeling that he had overvalued Mendel's article and that his own theories were better after all. Six years previously, he had first suggested that hereditary features were transmitted by vibrations rather than by discrete particles, and he had become so preoccupied with his own hypothesis that he preferred to dismiss all other explanations of evolution out of hand. He summarised Darwin in one line: 'Selection is a true phenomenon; but its function is to select, not to create.'[5] He was convinced that Darwin had got it wrong. He thought the London group who surrounded Galton and Pearson, and who seemed to do nothing but analyse data, were no better. After all, there was still a possibility that vibrating messages would show Mendelian ratios and that he had been right to get excited about discovering the manuscript. He claimed further support from mainland Europe where the prominent botanist Hugo de Vries was finding evidence that mutation could explain hereditary change; de Vries had even found evidence that one species could split into two. These big and sudden evolutionary changes were much more dramatic than anything Darwin had anticipated in his vision of a world of gradual change.

Through the 1890s, relations between Weldon and Bateson had become frosty, but the real target of Cambridge's distaste for UCL was more likely to have been Karl Pearson. One of Weldon's friends was a mathematics student called John Maynard Keynes, and he began to question the usefulness and truthfulness of statistics, especially to help understand the readily available data from economics. He was concerned by some of Pearson's conclusions about social issues and asked whether 'alcoholic homes exert in general an evil environmental influence upon children, and if they could be investigated only by experimental, not statistical, methods'.[6]

Many other people grappling with genetics also quarrelled with Pearson's emphasis on statistics. They were not happy about taking data from just one or two variables and assuming that 'all other things remain equal'. It was a problem for all scientific analysis: how to keep everything else constant out of the experiment, when just one thing is being considered. For instance, Keynes's example of the social problem of alcoholism was mixed up with the political ambitions of the temperance lobby on one hand and the eugenics lobby on the other. It could never be analysed statistically in isolation to give a clear picture. Such interpretation of data could never be entirely objective.

To show the importance of subjective interpretations, Keynes asked an apparently simple question: 'Do you expect it to rain today?' The answer depended on interpretation. Did the question mean: Is it *more* likely to rain than not? Is it *less* likely? Is it *as* likely? His point was that there were many alternative approaches to such an apparently simple question, but he did have a very simple answer: 'Take an umbrella.' This was a way of arguing about uncertainty that set Keynes apart from most of his friends. At the Saturday meetings of the Apostles, his youthful hero George Moore spent much time separating a person's innate goodness from his or her outer actions. Keynes asked whether without some Christian or Victorian standard, what sanction does an individual have to say what is good or not good? There was always variation in the passion of the human heart, in the pattern of life among communities, what Keynes called the 'powerful and valuable springs of feeling'.[7] He had outlined these concerns to the Apostles at one of their meetings in 1904 when he was looking for a scientific basis to morals, wanting to challenge Moore's Victorian utilitarian concept of goodness, but at that time he had not had the clarity or confidence to go public. It was to take him another sixteen years to publish these ideas in one of his most important books, *Treatise on Probability*, published in 1921.

Bateson shared Keynes's scepticism of statistics and saw no good coming from such an analysis. Instead he emphasised that mutation caused evolution which, in turn, could lead to a single species splitting into two. In his 1894 *Materials for the Study of Variation* he had explained this 'discontinuous evolution' in detail. *Materials* gave no place at all to environmental influences and concentrated instead on Bateson's own untested theory of evolution by vibrations. At that time, no one in Cambridge gave much attention to Darwin's ideas on natural selection, until Alfred Wallace was asked to review Bateson's *Materials*. The review was very critical of Bateson's work, especially his vibration theory for which there was no direct evidence. The day after that review appeared Wallace met with Weldon in London, and they talked about Bateson's dismissal of many of the ideas that Darwin had cherished. Wallace and Weldon went away intent on going public with more criticisms of Bateson's biased approach. In 1895, they both wrote critically of Bateson in the *Fortnightly Review*. This put what had been a strong friendship at Cambridge between Bateson and Weldon under considerable strain. Worse, Galton continued to support Bateson's book because he thought it added to the cause of his own polyhedron model. Galton had never really accepted Darwin's insistence that evolution proceeds at a slow and constant rate; he was always looking for support for the idea that evolution moved in steps.

Another battle in the war between Bloomsbury and Cambridge began harmlessly enough. One of the increasingly rare defenders of mainstream Victorian values, and a continuing supporter of Darwin in particular, was Sir William Thistleton-Dyer, the director of Kew Gardens between 1885 and 1905. Dyer thought that evolution led to distinct and stable species across all communities, and therefore agreed with Weldon about Bateson's neglect of outside influences on evolution. Dyer had made the same point in 1895 when he suggested that different coloured varieties of the ornamental flowering plant *Cineraria* were hybrids and not random mutants. He observed that the different colours formed gradually not suddenly. Whether this meant that they were not mutants was not very clear then. Dyer's support for gradualism attracted the wrath of Bateson in the correspondence columns of *Nature*, and Weldon soon joined in. Bateson took the criticism personally, and from then on his Cambridge group, which was studying mutations in what became Mendelian Genetics, felt they were distinct from the London school of statistical biometrics.

The differences between the two sides were clear. In London, Pearson, Weldon and Wallace continued to support ideas set out by Charles Darwin in *On the Origin of Species*. They believed that the major mechanism of evolutionary change was competition by natural selection. This was stimulated when environments changed and usually happened gradually within populations. In their view, this explained why evolutionary change was hard to spot and why large amounts of data from accurate observation were necessary before statistical analysis could show new groups emerging and old ones becoming extinct. In Cambridge, on the other hand, none of this was thought to be important.

For the Cambridge men, new species originated by sudden mutations at the genetic level and evolution was driven by processes deep inside the cells and at that time unknown.

Despite the attempts of Wallace and Weldon to bring Darwin's work into their discussions with biologists, the last few years of the nineteenth century had seen further decline in support for natural selection. Those who opposed Darwin's theories were encouraged by Bateson's arguments and the continuing absence of evidence for adaptation. Galton worried that he might die before the elusive agent of heredity was discovered. He was kept going by the popularity of Galton's Law which derived from his work on the pedigree of the basset hounds. In this he used new data to establish the distribution of each parent's contribution going back four and more generations. Pearson, overjoyed with such good data for analysis, described the patterns of inheritance for the white and yellow patches with a series of new equations. To cheer up Galton, Pearson incorporated these into a Happy New Year card. Galton's warm response gave all the London biometricians a new lease of life for the next battle with Cambridge.

That battle came in 1901, sooner than they expected. Bateson was the referee of a manuscript setting out Pearson's equations. Unfortunately, although the mathematics in the article may have been good, the biology was bad. Bateson admitted to not understanding mathematics, yet he took great delight in rejecting the manuscript. This angered the London group who decided to begin a journal of their own for such articles that crossed the boundaries of traditional disciplines. They called the journal *Biometrika,* which is still publishing high-quality work today.

The arguments about analysing data and distinguishing mutants continued unabated until Weldon's sudden death in 1906 at the age of forty-six. After that, some of the fighting spirit left Pearson, allowing Bateson to get on with supporting Mendel's ideas, while quietly keeping an eye out for evidence for his own vibration theory. Pearson was so disturbed by Weldon's death that he settled into a quieter way of life to concentrate on such projects as writing Galton's biography, but he also wanted to spend more time talking at conferences and promoting eugenics. Later, George Bernard Shaw, who admired Pearson's work, said: 'When I get my hand in sufficiently I think I will write *Karl Pearson, a Tragedy*. Can anything be done to rescue you from your professorship?'[8] Pearson was driven, at the expense of everything else, by the idea of achieving objectivity. He had discarded his earlier ideas as being too subjective. Shaw said that he had also felt the same temptation to study too many things and in the end settled for writing plays.

By 1907, Pearson was bemoaning his grand ambitions and pronouncing his career a failure: 'Twenty years hence a curve or a symbol will be called "Pearson's" & nothing more remembered of the toil of the years.'[9] Maybe this was the part of Pearson's life that Shaw regarded as tragic: the difficulty they both had in reconciling their socialist beliefs with their narcissistic enthusiasm for study. The same dilemma was to become a familiar feature in the lives of several Bloomsbury scientists and artists. Most of them were afraid of change, especially that represented by the new technology. Although Pearson wanted change, he was content to sulk with his mechanical Brunsviga calculating machine and to become even more distant until he eventually retired at seventy-seven.

There is no doubt that at the peak of his working life in the middle of the 1900s, Pearson did succeed in highlighting the importance of experiment in biology. Galton, meanwhile, had been productive in encouraging his students to gather large data sets for statistical analysis, and it led to an interest in patterns and clues about evolution from many different kinds of new sources.

**

Since his schooldays in Regent Square, Tansley had seen great social change in Bloomsbury, much of it brought about by scientific and technological advances. H. G. Wells had also been aware of this transition and by 1905 had become involved with several political groups connecting genetics, ecology and political science. The relationship between these three new fields was a central theme in his popular novels. He was aware that in the 1900s the political classes were increasingly turning towards the Fabian cause. Founded in 1884, the Fabians were a largely middle-class pressure group that advanced the principles of socialism. The Fabians followed the style of Fabius Maximus, who had been nicknamed 'the Delayer' in ancient Rome, slowly wearing down the enemy with peaceful tactics such as cutting off the supply of weapons.

The new group gained support from many influential Bloomsbury thinkers. To stimulate more ideas, its founders, Beatrice and Sydney Webb, started a monthly dining club called the Coefficients. Members included familiar intellectual characters such as the physiologist R. B. Haldane, Bertrand Russell, George Bernard Shaw, John Maynard Keynes and H. G. Wells. How small the social and professional circles were in Bloomsbury in those days! In both the arts and the sciences the same people asked the same questions. Most of those involved still had the view that science was leading to some kind of utopia, while still not accepting that religion and evolution might be quite separate matters.

By 1905, H. G. Wells was a famous figure throughout the English-speaking world. His main goal was to link the compatible methodologies of science and politics, and he was to do it slowly and peacefully through the Fabians. There seemed to be something inevitable about politics and science coming together as both were concerned with the same questions: population, social class and socialism. Wells explored these issues in his fiction. He was supported by his new friend, Lankester, who also believed that the mission of twentieth-century science was to bring together the common themes of genetics, ecology and political science. Wells's scientific training meant that, for him, everything had a purpose; if an answer was doubtful then he used his intelligence to manipulate a useful outcome. If there were no outcome, the analysis was discarded. For him, unlike for his friend Henry James, there was no beauty or art for its own sake, no time to reflect and ponder a scene for its atmosphere. The different perspectives of Wells and James caused the two men to engage in many heated discussions about writing and style. At first these arguments increased their admiration for one another's work, but their differences were so great that their friendship eventually had to break.

Wells struggled with his own anger about his background and about the privilege of those he emulated. His reputation, wit and confidence made him attractive to women, and he liked to play dangerously and have regular affairs. One was with Amber Reeves, the daughter of the director of the London School of Economics and Political Science. Moreover, in his 1909 novel *Ann Veronica*, he teased her publicly by writing: 'All the world about her seemed to be – how can one put it? – in wrappers, like a house when people leave it in the summer.'[10] Wells put immense pressure on his protagonist Ann, and his mistress Amber, to leave the rigid social habits that had emerged during Victoria's reign, to get away from a society still largely dominated by social discrimination and inequality of opportunity. The critic John St Loe Strachey wrote in *The Spectator* that the book was capable of spoiling the minds that read it. Predictably, Beatrice Webb, the rather prim founder of the Fabians, was quietly upset.

George Bernard Shaw also lectured to the Fabians, and, by 1903, he was convinced that science would be a major force in the future. Ever since the 1885 party when he joked about Weismann's tails, Shaw had been unhappy with Darwin's ideas and was searching for another way to explain evolution. Now he had one that linked to the ideas of the Fabian movement. In 1903, he wrote *Man and Superman*, which set out his campaign that eugenics and socialism could become entwined to give a promising future for distant generations. He also thought that a broader overview was needed not only to explain biological diversity but also the beauty of things such as Mozart's and Beethoven's music. Shaw and the Fabians had no doubt that the greatest biologist had been Darwin himself: it was his interpretation that Shaw didn't like. Yet, throughout the 1890s, no one else with an holistic approach like Darwin's had made any breakthroughs. Even the theory of natural selection fell into neglect.

However, Shaw also shared the idea, so popular at the time, that human destiny would be determined by natural creative impulses rather than by biological advances. It was what many thought of as the life force, what Henri Bergson the French philosopher, called 'creative evolution'. Men such as John Ruskin and Thomas Carlyle, Wagner and St Francis had tried to envision their own utopias, but Shaw had his own view: modern humans, he believed, could never obtain that state prophesised by Nietzsche. Instead, humans needed to assume that in the future our species would evolve into a particular kind of superman. This, in turn, would lead to the socialisation of the means of production and exchange, a positivist's dream of moralising the capitalist and an ethicist's dream of putting codes and lessons on a man as a harness is put on a horse.

Several other anti-science activists were busy in Edwardian Britain. In 1903, an anti-vivisectionist called Stephen Coleridge heard a rumour that the well-known UCL physiologist Ernest Starling was regularly performing experiments on a dog without even the most basic anaesthetic. The newspapers discovered that Professor Starling was indeed in charge of a research programme into diabetes and pancreatitis, and the work was being done by a wealthy student called William Bayliss. Denying the accusation, and having good evidence that the rumour was incorrect, Starling encouraged Bayliss to bring a libel case against Coleridge, and they won damages of £2,000.

In response, the anti-vivisectionists raised money for a statue of the so-called Brown Dog in Battersea Park, close to the dog's home. It had an inscription 'Men and Women of England – how long shall these things be?' The case and the statue became legendary for UCL students, proud that their college defended safe scientific research, and they regularly paid homage to the statue, but it was quietly taken down in 1907.

Another man who hated the idea of being 'crushed out' by science and machines was another Cambridge Apostle, E. M. Forster, even though he admitted that machines might help man 'get a new and perhaps a greater soul for the new condition'.[11] However, on the whole, science was all too much for Forster, and he feared its power. In 1908, he was spending increasing amounts of time in Bloomsbury teaching at the Working Men's College just around the corner from his gay friend Hugh Meredith's lodgings in Guilford Street. In one of his classes, he turned the novelty of early aeroplane flight into 'anthropomorphic theology up in the sky', an idea that became a short story, 'The Machine Stops', about a man escaping to the surface of the earth for a day. For Forster this was only a brief encounter with science, but an important one.

***

In 1908, there was an outstanding opportunity at UCL to see the influence of art and science, the one on the other. Gwen Darwin (later Raverat) was Charles Darwin's granddaughter and was learning her skills as a wood-engraver at the Slade School that year. At the same time, Agnes Arber was finishing off her postgraduate work in the botany department, on leaf development and morphology. How did talented women such as these make out as ambitious artists and scientists? Did art triumph over science? How were they bullied by men? There is little evidence easily available, but the playing field doesn't look level.

Although they worked in the same building, there is no record of their friendship there or through their early widowhoods later in Cambridge, but it would be surprising if they had not known one another. In her 1952 childhood memoir, *Period Piece*, Raverat wrote about a strong sense of guilt during her student days. Soon afterwards, there were comparable accounts of Arber's bad relationship with her supervisor, Professor Sir Albert Seward, where the lack of support for her research suggested that she, also, had been a victim of discrimination.

Later, in 1908, Adrian and Virginia Stephen invited Gwen to tea in their home at 29 Fitzroy Square. Coldly, they asked why Gwen was an artist, and, rather tensely, she replied, 'Because I have got to be.'[12] Later, Virginia reflected how they had jeered at the scientific thoroughness with which Gwen had taken to art. Gwen and her husband Jacques Raverat called themselves neo-pagans, enjoying an outdoor life with daisy chains, long country walks and talk of their young generation supporting Fabianism.

It was normal for women to react to bullying in this way, and it was common for much of their talent to be wasted. Less commonly, both these women, one a scientist, the other an artist, showed very similar creative talent that was not entirely wasted. From Raverat there are woodcuts and *Period Piece*. Agnes Arber was balancing nature and philosophy. She argued that the use of pictorial imagery in thinking is a fundamental need of the human mind: in biology you only really look when you draw. Both women gave particular attention to structure and function, with a view of nature through their own eyes and interpretations of its images in their own minds. By different means, they both found unique ways of sharing the information and understanding its beauty and meaning. Arber's 1953 legacy, *The Mind and the Eye*, is another classic memorial of how she so cleverly interpreted what she saw.

The Slade School at University College was a major centre of English art, and there was a natural rivalry between several groups of painters. A leader of one of these was Walter Sickert, who had been strongly influenced by the French impressionists and had moved away from trying to idealise the nature that he saw all around him. Artists were no longer recording something just because it existed; rather, they were being selective and using increasingly abstract styles. Sickert had been taught the science of colour by Lucien Pissarro, the son of the well-known French artist, who was living in London. They met at 19 Fitzroy Street on Saturday mornings in the summer of 1908 to work together.

It was in that part of Bloomsbury that the middle class from Kensington had invaded and introduced fashionable alternatives to the old traditions. However, other artists wanted to replace that romanticism with reality. Artists such as Spencer Gore, Charles Ginner and Harold Gilman attended Walter Sickert's weekend workshop and painted vulnerable women lying motionless on squalid iron bedsteads, stern landladies and gas stoves; they were sad observers of the social divide in English society. Social class was a major theme for the Camden Town group also. In 1911, the group elected sixteen members including Augustus John, Duncan Grant and Wyndham Lewis. The pictures painted by these men often brought together groups that traditionally avoided one another: for instance, working-class men and women could be seen associating with upper-middle-class intellectuals. The pictures were not the usual romantic idealisation but instead they were strongly unsettling. Separately it seemed as though something awful was always about to happen. These were expressions of the same fear of sexual pleasure that Marie Stopes was thinking about from a biological as well as a social angle.

The Camden Town group took subjects that were clearly part of the local lifestyles and painted them without regard to personal feelings. It was similar to the split in Lankester's world of biology: one side hanging on to the beauty of nature's complex environment, the other one measuring the details inside each cell. The Camden Town artists did not allow emotions to cloud their objective views of nature, and, much like the local scientists, they left their personal warmth for their family. The subjects were cold and withdrawn, making no effort to connect with their fellow humans. Their pictures of degenerate life so close to Bloomsbury teased the public's imagination in a similar way to H. G. Wells's *The Time*

Machine. They played on people's fear about where society might be going without the control of a Victorian moral structure.

When Lankester first read Wells's *The Time Machine*, he didn't know that the author had been one of the students whose work he had failed in the second-year examinations. Equally, few of Wells's readers were aware that Lankester had written a book about degeneration, though they were certainly fascinated by what sounded very different to the popular view of evolution: regression and decline. These ambiguities were to stimulate many of the forthcoming problems of the new century, such as what to do about homosexuality, dandyism, mysticism and naturism, all of which were coming into vogue as though manifestations of a degenerate society. Turning the African into the cultured European was considered progress, but degeneration suggested that things could go the other way as well. Was the loss of mental and physical control part of the evolutionary process? As Wells's traveller says in *The Time Machine*: 'the too-perfect security of the Upper-Worlders had led them to a slow movement of degeneration, to a general dwindling in size, strength and intelligence'.[13] But what was to be done?

Science and art were preoccupied with the same set of questions about human evolution. Since race and social class seemed to be mere stages in that change, maybe science could help it advance and prevent it going backwards. That was the intention of those who advocated eugenic research. Proponents of eugenics agreed that the ideas needed more thought in order to eliminate the obvious dangers. Wells and Lankester knew one another well by 1906 and must have talked through the arguments together.

Lankester, Wells and Fry were respectful of the part played by art in science and used that link to understand the whole of nature and the interactions of its internal and external parts. The new generation understood enough from the past to be open to the world and to express feelings about their position in it. One example was the new biology that was beginning to explain sex and inheritance, ecology and altruism. Young people of Virginia Stephen's generation were angry that their parents couldn't relax their Victorian yoke about that and many other issues.

These Bloomsbury pioneers were now a distinct group, without organisation but with plenty of influence on one another and on their generation. They were all looking forward to very different styles of living. The group included writers and artists as well as scientists – people regarded now as specialists – but, in Edwardian times, they were the intelligentsias without further subdivision. (The English have never felt comfortable describing such people as intellectuals.) As we shall see, the group of scientists and artists was growing and its influence was being recognised by social scientists, the artists of the Camden Town group as well as the more famous residents of Gordon Square. During the same decade, the group's identity became distinct from their more focused adversaries in Cambridge.

The success of science education encouraged an increasing number of people to challenge their religious belief. Looking further ahead, people such as Russell and Keynes were also asking whether the natural sciences, and biology in particular, might one day be able to explain everything scientifically. They understood that social and political affairs were often conducted irrationally and asked how scientific reasoning would ever be used in that kind of political sphere. The new popular movements of socialism and eugenics were about to put their fears and joys to the test.

7
The rise of eugenics, 1901–14

Despite the successes in opening up the natural sciences to more students and a wider public, Lankester and Wells knew that the slow progress in understanding evolution and genetics would make it harder for them to improve the lives of ordinary people. For statisticians such as Karl Pearson at the beginning of the new century there was a frustrating lack of numbers and measurements from biological investigations. So in 1901, when social scientists such as Beatrice Webb and Charles Booth began to base their case for social reform on their own new sets of numerical data, the Bloomsbury scientists became involved.

One data set had been published by Webb and Booth. They had monitored the wealth and living standards of people living in London. Their study allocated 13,600 streets to one of ten categories describing the residents' social status and showed the results on a colour-coded map. The undernourished and unemployed had dark colours and dominated the east of the city around the docks and the slums. This contrasted with the brightly coloured west side where there was green space and fresh air. In Bloomsbury the streets were in the middle of this spectrum and showed on the map as purple. The data filled seventeen volumes of *Life and Labour of the People of London*, and showed details of things like the higher birth rate in the dark-coloured streets in the east and the better health in the brighter coloured streets to the west. Pearson was delighted with the data, if not with the social conclusions of his statistical analysis.

Although Webb and Booth knew that the study would help their social and political cause, they were less certain about how their aims could be achieved. There was a strong tendency to intermarry within the same class, so they expected a marked difference in the richness of intelligence between the classes. Booth knew that life in London was hard, but he had no doubt that the lower levels of brainpower were 'hereditary to a very considerable extent', and that was the common impression at the time. It was believed that poverty and misery were largely genetical features of humanity and that social environment had an insignificant role. Increasingly, it appeared that biological science had a role to play in eradicating poverty and raising living standards for all people.

This was the main foundation of eugenics, an assumption that the progress of the human species could be controlled and improved through selective breeding. It was an idea developed from the eccentric enthusiasms of two men in particular, Francis Galton and Karl Pearson.

Convinced that Galton was right to want to reduce births, Pearson persisted with his own political vision to improve the genetic composition of human populations. He believed strongly that it could be achieved by encouraging statistically selected groups of men and women to reproduce. For him, at least, that was what the Men and Women's Club had been about. Others in Bloomsbury such as the many middle-class women, supported the eugenics movement because they saw it as a way to advance their suffragette cause. Eugenics seemed to offer them control over their reproductive roles. They also wanted to improve the quality of life for feckless working-class girls. One of the slogans was 'Fit women are carriers and regenerators of the race', though it wasn't clear whether they meant the 'human race', the Anglo-Saxon race or the British race.

Pearson was not bothered with these intricacies. He was content to use the suffragettes' support as promotion for his own different beliefs. He argued even harder that the individual was inferior to the group and that a person's feelings should be overridden by scientific objectivity. This view was strengthened in 1904 when Pearson's colleague Charles Spearman developed a powerful statistical tool that he called factor analysis. Factor analysis compared individuals' differences and allowed intelligence to be quantified, a tool that eugenicists found very useful.

Anxious to test Darwin's ideas of migration patterns for human communities, and being forever short of data to develop new statistical methods, Galton's and Pearson's theoretical ambitions attracted considerable attention, first from the academic community and then from the politicians. There was also public support to improve the quality of human populations and reduce the reproduction of undesirable human characteristics, plenty of discussion about the difficulties and dangers, and plenty of suspicion about giving scientists or politicians too much power, but there was never any clear agreement among politicians about how to select either the candidates or the methods.

Lankester, for one, was more than suspicious about many of the aims of these pioneers. He wrote articles in the newspapers demanding clarity about what was meant by some of the terms eugenicists threw around. He also pointed out that eugenicists confused education and inborn characters. If the cessation of selection led to racial degeneration, then it was in the richer sections of the community where the effects would be most obvious, not in the slums of the 'half-starved, struggling poor'.[1] Lankester was one of only a few people who protested at the eugenicists' ideas and at the programme of legislation that was threatening to restrict the breeding of the 'feeble-minded'. With his usual determination, he demanded that before too many people jumped to too many conclusions, more should be known about the origin of human characteristics, whether they were inherited or learnt. However, very few of the political leaders seemed to understand these concerns. To simplify the arguments, Lankester questioned what some of the terms meant. Terms such as 'racial quality' and 'improvement' meant different things to different people, and he called for a serious study of human races and their origin. However, no one answered his call. In frustration, Lankester made several criticisms of eugenics in his weekly *Daily Telegraph* column, but the articles seemed to have little effect on the public perception of those he called the 'intellectual eugenicists'.

In his articles, Lankester stressed that there was nothing unusual about congenital defects. They occurred in considerable numbers of species of animals and plants. It was part of the complex process of natural selection. He cautioned those who wanted to interfere with these processes, and he explained why different kinds of 'feeblemindedness' could occur in the offspring of parents of all social classes and wealth. He argued that if biologists had trouble defining terms and limits, then politicians would almost certainly make a mess of it. In his heart he knew that legislation was not going to work, and he felt frustrated by not being able to say why. Not knowing the full explanation of life's secrets was becoming dangerous as well as embarrassing.

In 1904, in order to help with their promotion campaign, Galton and his friends raised enough money to persuade the University of London to set up a eugenics record office in Bloomsbury at 50 Gower Street. The office would support 'the exact study of what may be called National Eugenics, the influences that are socially controllable, on which the status of the nation depends'.[2] Even Bernard Shaw was enthusiastic that there was no reasonable excuse for refusing to face the fact that nothing but a eugenic religion could save civilisation. The office was for research and worked closely with the separate, education-centred Eugenics Society.

On St Valentine's Day the following year, Galton addressed the membership of the society with a lecture he titled 'Restrictions on Marriage'. Restricting marriage was one way he envisaged controlling the health and intelligence of future populations. The meeting was packed with his supporters who warmly approved the idea. One of these was Dr Frederick Mott, the chief pathologist of London's asylums and a fellow of the Royal Society, who went on to write about how to segregate the unfit and check their reproduction. He called this 'improvement of the stock'.[3]

By then, Pearson had become a powerful figure and was well known to scientists and political organisations. His belief that statistical correlation could find meaning in large sets of data from individuals fitted with his proposition that every atom in the universe, and therefore every human being in the world, was different. This turned out to be the basis of Pearson's socialism that posited that the ultimate basis of knowledge was correlation, not causation. Even Galton found this a bit hard to take, especially without any hard evidence that every atom was different, but Pearson was not to be sidetracked, and when he addressed the Eugenics Society in 1904, he declared, 'The aim of eugenics is to represent each class or sect by its best specimens; that done, to leave them to work out their own civilisation in their own way.'[4] He also had his own clear ideas about which characteristics to favour: energy, ability, manliness, health and courteous disposition. He concluded that 'what nature does blindly, slowly and ruthlessly, man may do providently, quickly and kindly'.[5]

Pearson had been doing a very good job at recruiting supporters from the new middle classes of British society. He was helped by the strong surge of social change and the rapid scientific advances at the end of Victoria's long reign. The growth of opposition to his eugenics campaign was slow and quiet, but it lacked any kind of organisation. This was because it had come from its likely victims, from the working-class people themselves, and they had neither the education nor the resources to provide coherent opposition.

H. G. Wells waded into the eugenics debates in May 1904. In an address to the old Sociological Society, he directly criticised Galton's latest suggestion that bishops' sons should be encouraged to breed while those of criminals should not. Surely, he said, this was confusing nature with nurture, ignoring the social and psychological effects of poverty and social hardship. Wells had argued two years earlier in a lecture he called 'The Discovery of the Future' that the French philosopher Auguste Comte and the positivists were wrong when they said that European society was in transition between two phases of stability. Wells knew enough about geology to say that change was continuous and that the rate and subjects of change also varied. Science was revealing a much more complicated set of processes than Comte had expected. In response, Galton wrote a brief article in the *Daily Chronicle* under the headline: 'Our National Physique – Prospects of the British Race – Are We Degenerating?' Written just after the Royal Society had awarded him the Darwin Medal, it argued that 'a material improvement in our British breed is not so Utopian an object as it may seem'.[6]

If the article had been written specifically to annoy Wells and Lankester, then it certainly worked. Their fury was just about balanced by their admiration of Galton as a pluralist, one whose interests also included fingerprints, photography and meteorology. However, their arguments failed to hit their target with much force. They contended that Galton had mixed up two different meanings of 'degeneration' and had compared them to even more confused meanings of the word 'race'. Wells wrote, 'I believe … it is in the sterilisation of failure, and not in the selection of successes for breeding, that the possibility of an improvement of the human stock lies.'[7] He had hinted at controlling genetic illnesses a decade earlier with degenerate man-creatures in *The Time Machine* and had thought through some of the social consequences. Wells was a man of conflicts, pitting art and science, men and women, love and hate against each other. He was a member of the Eugenics Society, but he also argued against its objectives; he challenged the political right and left with equal force.

There was not a lot known about genetics or the biology of human populations for any of these men to take as sound background for their political agenda. Darwin and Weismann had provided the most useful data, and T. H. Huxley had been its most inspiring orator and educator, but the knowledge available then was only very basic. The other protagonists in the debates about population, race and mental illness knew even less, and the pressure on them to make clear recommendations to the politicians was immense. Lankester and Wells recognised this, and often said so, but strong social pressure demanded some kind of action to reduce the needs for the dreaded welfare institutions.

Another kind of objection to eugenics came from the Christian writer on evolution, Benjamin Kidd. He was 'appalled' by Pearson's support of Galton's political pronouncements, especially his challenge to the human rights of the individual. Should humanity retain the law of ancient Greece and rank each person as subordinate to the claims of the state? Kidd wrote about 'equality of opportunity' rather than the same social level for all members of a group, let alone the whole of society. This topic of the genetics of whole populations was a recurring theme in his work, whether and how kindness and cooperation kept individuals together.

More opposition came from Dr Henry Maudsley, an acquaintance of Charles Darwin, and a psychiatrist. Maudsley was sceptical of Galton's argument that talent and character could be inherited. Instead, he favoured the theory that families like the Darwins were successful because they were privileged to have grown and developed in a favourable environment. Similarly, he was satisfied that William Shakespeare stood out from undistinguished relatives, just like many other prominent individuals who had mediocre relatives and even some with a mental disorder. Maudsley found no reason to favour nature more than nurture, and H. G. Wells agreed that Galton's classical work was premature. Another critic of Galton and Pearson was a doctor interested in sexual behaviour, Havelock Ellis. Ellis did not like Galton's comparison between animal breeding and human eugenics. We may be good at breeding other species like cows and horses for particular purposes, but then he asked whether we can be sure they would not rise up and control us. He thought it would require a race of supermen to successfully breed new human varieties and to keep them chained up in their stalls.

However, those with doubts about eugenics were in the minority. Many other establishment scientists were easily recruited to support the eugenic cause. Galton canvassed fellows of the Royal Society to find evidence of the heredity of high ability and wrote his views in *Noteworthy Families*. In this book he argued that noteworthiness diminishes rapidly as the distance of kinship to the Royal Society fellowship increases. He took this to mean that 'able fathers produce able children in a much larger proportion than the generality'.[8] He was pleased with the argument and used it to draft his eugenic certificates with 'memories of the useful achievements of the kinsfolk'.[9] These were to be given to selected applicants by the Eugenics Society. The proposal was announced by Lord Rosebery, Chancellor of the Exchequer, when he opened the new building for the Eugenics Records Office in 1905. Galton was delighted by this and thought it represented the peak of his own achievements.

Bertrand Russell was particularly enthused by the ideas surrounding eugenics and was tempted into the debate by its ethical importance. He first published on eugenics from an academic perspective. He tried to test Pearson's statistical reasoning and the principles of Galton's law of ancestral heredity, but his mathematics did not reach any clear conclusion. Instead he started to speak publicly about their ideas, suggesting that the state should make direct payments to 'desirable' parents while withholding funding from 'undesirable' ones. He was concerned about the 'dangers' of the so-called 'differential birth rate': the Edwardian concern that the poorer sections of society reproduced much faster than the wealthier. It was this that conditioned Russell's understanding of parental desirability.

Russell's views on eugenics developed from his earlier theoretical ideas as an academic mathematician. He shared both Galton's and Pearson's optimism about the opportunities that eugenics offered for the betterment of humanity, and he genuinely believed, with many millions of other people, that eugenics could improve society. However, this view was increasingly outweighed by fear about the uses to which eugenics would be put. Eugenics means 'good in stock, hereditarily endowed with noble qualities', and Russell wanted to extend this to the overall condition of humanity. He tried to encourage the 'fit' to breed more and to do so only with other 'fit' individuals. He also advocated 'negative' eugenic measures that aimed for a decrease in the fertility rate among the 'unfit' by either separating so-called defectives and undesirables from society or preventing them breeding altogether through medical sterilisation. Eugenics linked three strands of science: Malthus's warning of an unsustainable human population, Galton's studies of population statistics, and a different set of political attempts to regulate population based on ideas from population genetics.

To its supporters, eugenics was a progressive, rational and science-based project. Political and social conservatives did indeed support eugenic proposals, but so did political progressives for whom eugenics offered a solution to outdated and unethical practices. In their view, eugenics would help to improve the intellectual capacity of society and diminish the number of people suffering mental disabilities. Russell believed that it was scientifically possible to intervene in the reproduction of human populations with the specific goal of improving the biological make-up of future generations. He wrote to his fiancée, Alys Pearsall-Smith, about the idea of issuing marriage suitability certificates. Some of the Russell and Pearsall-Smith families disapproved of their union and pointed out to them, by way of warning, that there was a prevalence of mental illness among both families' ancestors, and that it would therefore be dangerous for them to have children of their own. It was to be many years before Russell changed his mind about eugenics, after realising how such practices were liable to political and personal manipulation.

Galton, though, saw eugenics as protecting the common good: 'Society should in many cases actually prevent the act of procreating and may without any regard for rank and under certain circumstances

Figure 7.1   Bertrand Russell (1907)

have recourse to castration.'[10] Pearson soon assumed the role of his 'statistical heir' and started to write a three-volume biography of his hero. Pearson firmly believed the theories of mathematical physics were going to unite 'the wonderful tales of the development of life'.[11] This was going to lead to new discoveries that would take humanity forward to an improved level of civilisation. Charles Darwin's *On the Origin of Species* was not featured in the biography; instead there was Galton and Darwin's grandfather, Erasmus.

Pearson was appointed as the first Galton professor of eugenics at UCL soon after Galton's funeral in 1911. The appointment encouraged some newspapers to argue that Pearson's work offered proof that racial degeneration was a threat to the Empire. Improved health allowed genetically inferior people to survive and stopped natural selection by propagating unfitness. In an article entitled 'Darwinism, Medical Progress and Parentage', Pearson repeated this theme: 'The right to live does not connote the right of each man to reproduce his kind. As we lessen the stringency of natural selection, and more and more of the weaklings and unfit survive, we must increase the standard, mental and physical, of parentage.'[12] Pearson expected life to be a struggle and welcomed the part of that struggle that led to survival by fitness. 'National progress depends on racial fitness, and the supreme test of this fitness is war. When wars cease mankind will no longer progress for there will be nothing to check the fertility of inferior stock.'[13] The 1912 International Congress of Eugenics opened in London at the Hotel Cecil on the Victoria Embankment. The Cecil was the largest hotel in Europe and had a powerful image that was to symbolise the status of the new field of eugenics. This was also the style of the president of the congress, Major Leonard Darwin, the fourth son of Charles Darwin and the only one not to become a creative scientist. Leonard Darwin had served for twenty years in the Royal Engineers and was Liberal Unionist MP for Lichfield from 1892 to 1895. He was then active in the Unionist Free Food League where he developed an interest in eugenics. At the congress, Darwin had many powerful deputies, including the vice-chancellor of the University of London, the chancellor of Stanford University, the lord chief justice, and senior eugenicists from France Italy and Germany. Winston Churchill and August Weismann, father of the germ-line theory of biological continuity, were also in attendance. The former prime minister, Arthur Balfour, welcomed foreign friends and guests. *Nature* wrote that the 'series of brilliant entertainments organised by the hospitality committee, under the secretary-ship of Mrs Alec Tweedie, was a bait which attracted many' of the 750 participants.[14] A special train took the participants to the conference banquet in the Kent countryside. With such a line-up of dignitaries, it was clear that something important was happening.

In his opening speech, Balfour set out the two aims of the congress. The first was 'to convince the public that eugenics is one of the greatest and most pressing necessities of our age'.[15] The second was to convince them that eugenics was also 'one of the most difficult and complex tasks science had ever undertaken'. He explained some of the difficulties:

> We say that the fit survive. But all that means is that those who survive are fit. A basic premise of eugenics that 'the biologically fit are diminishing in number through the diminution of the birth rate' must be wrong by the doctrine of natural selection. If families of the professional class were so small that it is impossible for them to keep up their numbers, they are biologically unfit for this reason.[16]

Once again, the London Asylum psychiatrist, Dr Mott, was one of the speakers and praised the proposals laid out by the Eugenics Society. He argued that mental illness could be controlled by applying eugenics. Mott described a correlation between the incidence of tuberculosis, pauperism and insanity and was convinced that tuberculosis was Nature's way of eliminating the unfit. Tuberculosis was, he said, natural selection in action. Mutation, natural selection and degeneration held the key to a happier future. 'Degenerate stocks generally contain feeble-minded of all grades, the majority of which will not die out, but propagate freely.'[17] Because the female side is free from taint and almost every number of the male stock unsound, tuberculosis could be controlled by sterilising the male paupers and insane.

In his closing address, Leonard Darwin emphasised the importance of nature by quoting Galton: 'If cattle breeders were urged to delay stock improvement because the laws of heredity were incompletely understood, they would simply laugh at us.'[18] There were roars of approval when he went on to talk about other themes such as involuntary segregation, sterilisation and racial intolerance.

The congress laid the way for the 1913 Mental Deficiency Act. The law allowed particular institutions to detain unfit citizens such as idiots, imbeciles, the feeble-minded and moral defectives. In 1920, there were 10,000 'mental defectives' in institutions in England and Wales. At the peak, 60,000 people were detained in Britain and another 43,000 supervised by control orders. Many who supported this policy believed the traits of unfit categories were inherited and looked to biology to enable some rapid cure. This was one of many ways in which the eugenicists thought they could control the growth of human populations. They also wanted to plan the structure of human society and its behaviour. There were voluntary eugenic mating societies in India, in Jewish communities and in the English aristocracy. To many, eugenic ideas seemed to come very easily. Walking beside the river in 1915, Virginia Woolf met 'a long line of imbeciles'. She wrote in her journal: 'It was perfectly horrible. They should certainly be killed.'[19]

***

Three years before the 1912 congress, Leonard Darwin had met Ronald Fisher, a brilliant young biology undergraduate at Cambridge. His father was a partner in a firm of London art auctioneers and owned a big house in Hampstead. The subsequent discussions between Darwin and Fisher made a strong impact on some members of the Darwin family. The connection provided Ronald with a lively intellectual background. His early happiness was rudely broken, however, by his mother's death and his father's financial ruin. He was still at Harrow School when these things happened, and he responded by burying himself in work and right-wing politics. He won a scholarship for mathematics at Cambridge, and Harrow rewarded him with a prize of Darwin's complete works.

Fisher helped John Maynard Keynes and Leonard's brother, Horace Darwin, establish a eugenics society in Cambridge. The first meeting was held in 1910 and was intended to be addressed by leaders of mainstream eugenics organisations, such as the Eugenics Education Society. However, the eager young Fisher beat them to it and talked himself. In his short address, Fisher cautioned both sides of the old Galton–Bloomsbury Bateson–Cambridge divide against explaining evolution without direct evidence. He pointed out that Mendel had published unusually accurate results, conforming almost too much to the expected ratios and argued that this was not science but the opposite: having an idea and selecting data to support it. A scientist used data to test the idea and try to show it to be wrong. Biometricians could be ambiguous, Fisher said, and 'could squeeze the truth out of the most inferior data'.[20] But the dispute with the Bloomsbury statisticians was becoming stale by then, four years after Raphael Weldon's death. During that first year at Cambridge in 1910, Fisher listened attentively to the professor of biology, William Bateson, the man who had invented the word 'genetics' five years earlier, and he didn't agree with a lot that was being said.

Bateson, however, was from an older school of biology and shared Lankester's broader outlook, the view that differences are necessary for progress. If populations were homogenous, their civilisation stopped. Bateson often explained that the common desire for all men to be equal, and to have equal rights, was contrary to the biological facts of nature. It was because of mutational novelties that progress occurred and, without that variation, a grey and amorphous stasis set in. In a socialist society, colourful and shapely people could be seen as a disturbing nuisance. Like socialists, the eugenicists advocated homogenous groups, just as socialists wanted level aspirations. For these reasons, Bateson argued that biologists must consider class distinction an essential attribute of any society. Without it, evolution would stop.

In 1913, Fisher told an audience of biometric research workers in Gower Street that if the evolutionary changes came about by mutation, or any mechanism other than natural selection, the sums showed that the resulting organisms would not survive. Fisher's own statistical analysis of the small amount of data available on Mendel's crosses showed that change was as you would expect if evolution came about by natural selection. Natural mutations were too rare and random, and the few offspring that did survive could not account for the selection of new adaptations. Fisher's idea of monitoring the frequency of mutation came out of the blue and was far ahead of its time. It led to a brilliant proof of Darwin's theory but it got lost in history, perhaps because it was so theoretical. Something practical was needed to attract the attention of non-specialists. Thus Darwin's theory went back into the shadows and was lost from view entirely during the First World War.

Later, when he was professor of eugenics at UCL, Fisher reflected on his role as arbiter:

> Each generation, perhaps, found in Mendel's paper only what it expected to find; in the first period a repetition of the hybridization results commonly reported, in the second a discovery in inheritance supposedly difficult to reconcile with continuous evolution. Each generation, therefore, ignored what did not confirm its own expectations.[21]

Fisher's generation of young men were more energetic and strong-minded than most, expecting measurements to be involved in all scientific problems. What did happen was that physics and chemistry were to be heavily involved in the forthcoming war. People had no time for the 'softer' bits of evolutionary biology. The specialists who recognised and described new species were still left out, and the new specialists who looked at the environment were not even considered.

Ray Lankester was powerless to intervene in 1913, despite his words of reason. He continued to use his 'Easy Chair' column in the *Daily Telegraph* to oppose eugenics, but his arguments were losing their vigour and did not get through to many readers. He said that the legislators had never really thought through what the terms 'imbecile' and 'feeble-minded' meant, let alone how to recognise them. They did not show a complete understanding of the eugenicists' view of biology.

More ominously, there were already signs that politically inspired medical men were seeking practical information from scientists. In 1910, Bateson was director of the John Innes Horticultural Institute where he was studying the genetics of *Antirrhinum*. He became friendly with a visiting specialist in genetics, Erwin Baur from Germany. Baur had been a psychiatrist, and that perpetual contact with the mentally ill had repelled him deeply. It was said that he was inclined instinctively towards all that was healthy; ten years later he published a book called *Human Heredity* that inspired Adolf Hitler when writing *Mein Kampf*.

When Bateson started to defend the bohemian spirit of a place such as Bloomsbury or Cambridge, Baur disagreed, expecting that such bohemians were destined to perish. Instead, Bateson wanted them to be supported by the state. He saw them as the salt of the earth, without whom the savour of life would be flat and wearisome. The great strength of the modern society was its polymorphism, and Bateson wanted all the differences to be together in a coordinated community. Through these years, William Bateson had learnt enough about nature to realise that playing around with complex genetical systems was going to be very dangerous.

8
Old habits die hard, 1901–14

Just before the First World War, it was easy for educated people to become involved with science in some way, using its technology perhaps, or gathering data for analysis by reading and talking about a particular theory. These were the ways Lankester used to inform the educated public about evolution and its consequences. For many of the younger generation, evolution raised the question of whether science could become part of their working lives. If they were attracted to science, what did they have to do to become scientists? From his place at the cutting edge of research and learning in the life sciences, Lankester came across three kinds of response to this question. These responses gave different levels of optimism about the rise of the professional scientist.

The first response was from a group of scientists and artists in their prime who enjoyed a challenge and argued for change. They were men such as Wells and Fry who were still being led by an older generation of people and institutions fixed in old ways of work and resistant to change. The old administrators at the British Museum were examples of this, men who had not been trained as scientists let alone as managers but who found themselves having to cope with the genesis of strange new scientific practices and strong-minded young specialists. It was common for these old regimes to be challenged by the reforming schemes of people such as Lankester and his contemporaries.

Of course, the reluctance to change was felt by the young men themselves, and that kind formed the second group. In families such as the Darwins, the Stephens and the Stracheys, a topic for conversation at their dinner tables was whether science would be something to consider as a career. Few of these men – for they were mostly men – wanted to break out of the familiar mould made for them by family and society; those who did want to go their own way did not know how. Moreover, professional scientists of this second group were not paid well. Quietly, they did the routine hard work in the back room.

The third and most adaptable group was made up of ambitious outsiders. These were people who found ways to live free from earlier historical customs and to whom science presented no such problems. They had fallen in love with science at an early age, often helped by their parents to learn basic science, and they saw it as an escape from working-class drudgery. In Bloomsbury there were rising stars like Arthur Tansley and Marie Stopes whose families had little money but gave plenty of inspiration and encouragement. They all had inventiveness and plans to implement their new ideas, though their strategies were unclear and some of them met unexpected difficulties and opposition. Most of the pioneers in my story were this kind of person.

Lankester had never found it easy to take orders. It was unsurprising, therefore, that when he served as director of the British Museum (Natural History) from 1898 to 1907, he clashed with the conservative and still-powerful trustees. Lankester and the trustees had many disagreements, including one in particular about the domination of one social class over another. As we shall see, Lankester won some of the battles but lost the war. Significantly, he was a low-salaried employee with no family status or capital. In contrast, the three senior trustees were the Archbishop of Canterbury, the Speaker of the House of Commons and the Lord Chancellor.

In 1898, Sir Edward Maunde Thompson, a distinguished palaeographer, became secretary to the trustees. Thompson planned to govern the whole of the British Museum, despite the move of all the natural-history specimens to the new building in South Kensington. The new building was still called the British Museum (Natural History), and control of the staff remained with Thompson, despite 129 signatures from fellows of the Royal Society supporting Lankester's appointment as director and the museum's autonomy. The archbishop was reluctant to support Lankester and told him so. 'I hear you are a very quarrelsome man', he said.[1] However, after pressure from the science lobby, he withdrew his objections, and the trustees offered Lankester the job.

At the British Museum in Bloomsbury, changing management styles revealed some cases of corruption that had grown out of old Victorian practices. Lankester found himself to be a victim of this corruption. He knew that the old gentlemen administrators were serious barriers to progress and throughout his eight years as director he was involved in many principled battles with these authorities.

Lankester believed deeply that the role of the museum was to obtain new knowledge of natural history through research. He wanted this research to be of high quality and to be aided by young research assistants. He respected research immensely and wanted to invest more in it than in people. This meant that scholarships were to be awarded to projects rather than people, as was common in Germany, where investment in knowledge was much higher than in any other country in Europe.

An example of Lankester's struggle with authority is the conflict that took place over the museum's insect collection. One of the trustees was an amateur entomologist and offered to sell his fine butterfly collection to the museum. There was a strong argument that such favours could prejudice the integrity of collections at the museum. Curators preferred to select the specimens themselves and the museum collections were supposed to reflect the evolution of the whole animal and plant kingdom not just the species that were the most attractive. In evolution, slugs are as important as ostriches. When Lankester refused to buy the butterfly collection, a huge row erupted at the next trustees' meeting.

After a formal enquiry into affairs such as this, the trustees decided that Lankester was neglecting the administration of the museum to pursue his own research. Once again, this was a clash of principle: the trustees wanted efficient and stable management, whereas the director wanted to educate all visitors in the wonders of evolution by natural selection. The trustees were less interested in the improvement of the exhibits or their increased popularity with the public. For example, they did not care about the gift of the giant *Diplodocus* from Andrew Carnegie despite the fact that dinosaurs were popular exhibits and new acquisitions attracted record visitors to the museum. On 12 November 1902, *Punch* magazine carried a cartoon of the *Diplodocus* saying to Lankester in an American accent, 'Wal! If he ain't a daisy!! Quite'n interesting specimen of the British professor! Carnegie'll just have to send a cast o' him over to the States right away.'

Another of Lankester's projects was an attempt to designate parts of Africa as national parkland. The aim was to protect large mammals from hunting and to search for new species. Several new mammals were discovered inside reserves, such as the mountain gorilla and the pygmy chimpanzee, but the most popular was a new species of okapi that Lankester described in 1901 and announced at the museum to great public interest. His *Okapi johnstoni* was a close relative of an extinct giraffe, with curious outer teeth. They had a chisel-like crown with a deep vertical slit and were on the lower jaw beside the eight front teeth. At the time of their first display at the museum, the Barnum & Bailey circus was in town, and Lankester asked some of the performers from Africa to come and look at the exhibits. They cried with surprise when they recognised the familiar giraffe-like creatures: 'Okapi! Okapi!' *Punch* published another cartoon showing Lankester riding on the creature's back: 'A ray of light on darkest Africa.'

This popularisation of science did not go down well with the conservative trustees. They had now been clashing with Lankester for more than five years, and there was no sign of Lankester giving way. It is doubtful whether the differences between the trustees and Lankester were just about the balance of management and science, for the personalities of the trustees' secretary, Maunde Thompson, and Lankester were very different; the former organised and introspective, the latter spontaneous and open. It is likely there was a deeper cause to the arguments: social class.

In 1906, the year of his presidency of the British Association for the Advancement of Science, Lankester's problems with the museum trustees got worse. New regulations had brought forward his retirement, and he was offered a pension of £300 a year; a derisory sum. In characteristic style, Lankester wrote a long account of his problems with the trustees in *The Times*, where an editorial strongly supported his case. The Archbishop of Canterbury became involved, as did the prime minister. The king, on advice of the prime minister, gave Lankester a knighthood. Yet, still the trustees held their ground and, in 1907, Lankester resigned from the museum. Some have said that the incident of his arrest in Piccadilly eleven years earlier had stained Lankester's character beyond redemption. In defence of the trustees, the archbishop wrote to Lankester saying that he knew nothing of a conspiracy: 'I certainly have never heard of the calumnies which you think must have been somewhere astir to the discredit of your personal character. They have not reached my ears.'[2]

Lankester responded to all the setbacks and delays by searching with even greater tenacity for evidence for the mechanisms of evolution. Realising that there was a lot of public interest in human evolution, he set about writing another popular book called *The Kingdom of Man* in which he suggested that humans emerged during Pliocene or Miocene times, around 10 million years ago, with the mental ability to make tools from material like the flint stones of the Sussex chalk.

Although the class of gentlemen scientists was slowly disappearing, Sir Edward Maunde Thompson's victory in the battle with Lankester showed the upper middle class were not done for yet. Another gentleman amateur scientist, Sir Arthur Conan Doyle, was said to have noted a femur and some other bones in a pit at Piltdown in Sussex before 1908. Doyle was a doctor and was always pleased to share his walks with his neighbour Charles Dawson, a local lawyer interested in collecting fossils. Together they visited one of the nearby quarries from which dinosaur bones were occasionally being unearthed. Dawson often found well-preserved specimens and intended to present the best specimens to the keeper of geology at the Natural History Museum, Arthur Woodward. In 1906, another neighbour, Cecil Wray, returned from Borneo with a collection of primate jaws and skulls, including some from the orangutan. Wray joined with Doyle and Dawson in neighbourly talks and expeditions to the quarries.

Doyle, Dawson and Wray were well known and highly respected, but did they have a say about how their discoveries were to be studied and analysed? Or did their finds just disappear into the laboratories of the salaried professional scientists? Lankester, meanwhile, got on with publicising this academic work. There was outstanding interest in the fossils from Piltdown. Some people used them to claim the site of human origin in Britain. If such a thing could be proved, it would be a triumph for nationalists, some of whom thought it would give greater legitimacy to King Edward and the Empire. In any case, Britain could finally match other European records of early human remains. In 1907, German palaeontologists had described *H. heidelbergensis* from Mauer then from Steinheim an der Murr, and the same species was found in France, Italy and Greece. A year later, in France, Professor Boule re-examined Neanderthal man and reclaimed its important place in human evolution.

Lankester was excited. For him, the nationalist argument didn't matter. The project itself was a good thing for science. He remarked that, in the past 2,000 years, learned men of Europe had debated whether this or that place was the site of ancient Troy, or whether there was ever such a place at all. He thought it was enough to inspire hope and belief in experiment.

Ray Lankester and H. G. Wells became close friends after they met socially in the Majolica Restaurant in South Kensington. Now, nine years later, they were members of the same club and often talked into the small hours of the morning about 'settling the affairs of heaven and earth'. Wells was Lankester's guest at the annual dinner of the Royal Society, and Lankester was invited as Wells's guest to the Omar Khayyam Club at Frascati's. In 1911, Wells visited the Natural History Museum and Lankester went to the Wells's home in Essex for weekends. This was when

Wells wrote his novel *Marriage* with its description of Lankester as one of the characters, Sir Roderick Dover.

Marriage continued to elude Lankester. In 1912, he had become friendly with the ballerina Anna Pavlova, whom he admired for the self-discipline and control she imposed on her own daily routine and on her relationships with others. Lankester must have worried about whether he could force himself into such a regime. He used to visit her at home in Hampstead, and he encouraged her to take on several roles for the London ballet seasons, but in 1914 she had to return to Russia. At the farewell dinner given in her honour by the Sadler's Wells company, Lankester gave the principal tribute: 'She dances as only a supreme artist can, the creative thought and conception of a mind of the rarest beauty and of the highest poetic quality.'[3] Pavlova turned and gave him a big hug in front of the whole assembly and was forever fond of her ageing professor. So much had slipped through Lankester's fingers: the search for agents of heredity, the British Museum, and now Anna Pavlova. At least he kept his scientific integrity.

Pleased to be free from institutional interference, Lankester satisfied his yearning to teach science through his weekly 'Easy Chair' column in the *Daily Telegraph*. The articles became an important part of the newspaper. They were lucid and free of jargon, and they always respected the reader's intelligence. Unlike so much other writing about science, they did not talk down to the readers. Like his lectures, they were both informative and entertaining. They had many clever drawings mixed with a sense of marvel about the beauty of nature and the ability of science to assuage our fears of the unknown.

The social mix of English society was changing, and the new scientists did not fit easily into any of the new structures. For example, many people felt the need for a very different attitude to deal with the challenges of the ethical gaps that evolutionary biology and technology had left unfilled. Was science and scientific development part of this daily need for guidance and do the conversations at the scientists' dinner tables rehearse some responses? What do the scientists say about the origin of life and the value of altruism? How do we treat our neighbours? Although Bloomsbury was atypical, with its freethinking network of intellectuals and artists, it did have an unusual responsibility to at least attempt to answer these questions.

***

In the spring of 1904, Sir Leslie Stephen died. Only a few weeks later, his four children, along with two of their servants from the old house in Kensington, moved into one of the houses on the east side of Gordon Square. The house served as a kind of laboratory for their own experiments into the future. They had much to look forward to, much to talk about and try out for themselves; they had the confidence and the money to do what they wished. Despite its leafy squares and Georgian terraces, Bloomsbury was not fashionable because its three big railway stations made the streets dirty and the cheap hotels attracted lonely souls. This, mixed with the university and the British Museum, gave the area a strong bohemian character, well suited to the Stephens's new mission.

The four orphans were excited by what their home had to offer. The house had six storeys, a kitchen and scullery in the basement and servants' rooms in the attic. The four siblings each had their own bedrooms. The reception room on the first floor was where their guests were entertained. The extra space gave them room to think.

The eldest of the Stephen children was Thoby. He had just left Cambridge to read for the Bar, though all he really wanted was to see more of his college friends. His sisters Vanessa and Virginia were in their early twenties, and their young brother Adrian just twenty-one when they moved into Gordon Square. Thursday evening was visiting time at the house. These occasions soon established themselves as major social and intellectual events in the Stephens's lives and those of many of their friends. It was 'at home' in the Stephen house that Leonard Woolf, John Maynard Keynes, Gwen Darwin, Bertrand Russell and Roger Fry met their London counterparts, the Stephens and the Stracheys. Ray Lankester, Arthur Tansley and Marie Stopes, however, were outside this group and were never invited to join in the parties or other insider meetings at Gordon Square. Although many of their interests and responsibilities overlapped and did bring them together, the Cambridge men and their women friends were separate from those based around UCL and the British Museum. Their art and science interacted in many ways at many times, but something else encouraged them to be separate. One group was inside the high London culture; the other remained outside.

Instead of family portraits hanging on the drawing-room wall there was a big painting by Augustus John. Virginia would 'talk egotistically and excitedly' about her own affairs, proving to herself and her friends that nothing was taboo. Scientific talk about the biology of reproduction had made the subject of sex acceptable. On one infamous occasion, the door opened suddenly, and the long and sinister figure of Lytton Strachey

Figure 8.1 Virginia Woolf (1902)

stood on the threshold. He pointed a finger at a stain on Vanessa's white dress.

> 'Semen?' he said. Can one really say it? I thought & we burst out laughing. With that one word all barriers of reticence and reserve went down. A flood of the sacred fluid seemed to overwhelm us. Sex permeated our conversation. The word bugger was never far from our lips. We discussed copulation with the same excitement and openness that we had discussed the nature of good. It is strange to think how reticent, how reserved we had been for so long.[4]

True or not, the incident demonstrates a growth out of immaturity among these young men and women, as well as their excitement about breaking taboos. They were beginning to understand that being independent and responsible involved managing their own lives, deciding for themselves how they were going to relate to one another and how they were going to be seen by other people. This realisation had an impact on most of those in the group. Virginia began to write fictional expressions of what she felt about her fellow travellers, and of her own isolation. This became the basis of her first novel *The Voyage Out*, completed in 1913.

Lytton Strachey was one of Virginia's closest friends. He was also struggling to find a direction for his life and was coming to understand the power of the Victorian conventions with which they were raised: 'There are three classes of human being, the rich, the poor and the intelligent. When the poor are serious, they are religious. When the intelligent are serious they are artists, but the rich are never serious at all.'[5] Perhaps, here, Lytton was thinking aloud about which group he fitted into himself. He realised that, despite parental pressure, many of his friends were not going to do the work necessary to join the new professions of science and technology; and he was right, their wealth made serious work unnecessary.

Most of this group were members of the Cambridge Apostles. As such, not only did they expect to behave and live honestly but some had brashly revealed their own feelings about human sexuality. Their eagerness to pursue these instinctual feelings was further encouraged by their conversations about the medical sexologist Havelock Ellis and their speculations about unconscious feelings on sexuality recognised by Freud. Oscar Wilde's trial and subsequent death had shocked many young gay men and reminded them about the dangers and public scandal of breaking the law. Those who had moved into Gordon Square knew it was best to keep their sexual adventures to themselves.

Roger Fry and Lytton Strachey were among those who asked what light the new discoveries in genetic science could shed on their sexuality. New arguments based on genetic evidence boosted their confidence in the normality of their own homosexuality. With no concrete explanations for the biology of homosexuality, however, they limited themselves to enthusing to their friends about the style and content of Charles Darwin's books and how they took his use of scientific methods to understand the world.

Young women from the middle classes felt a different kind of restriction. Most were still expected to spend their leisure time at home. On the few occasions when they went out in the evening, they were supposed to be chaperoned, and any kind of familiar exchanges with men of the same age were unusual. No wonder different lifestyles were cherished so much at 46 Gordon Square.

All four of the Stephens and many of their friends were interested in science as well as art, though Roger Fry and David Garnett were the only ones who had received any formal scientific training in biology. Garnett had studied botany and zoology at H. G. Wells's college in South Kensington and also went on to write novels. Some of the Strachey children had continued the interest in science that their father had tried to nurture, but their success was hindered in two ways. First, their rebelliousness and confidence that they were different from anyone else meant that they were not going to conform to the civil-service rules that had controlled their father's life in India, nor were they going to have anything to do with large institutions. Professional scientists usually worked in groups within institutions. They were chosen by a system of selection on merit. The Bloomsbury insiders, by contrast, were proud of their independence and of their difference from all the others. They had enough money and influence to run their own lives. Second, science did not offer them the opportunity to follow the kind of career they had grown to expect. Apart from medicine, few areas in the scientific field offered wealth or social status.

It would have been hard for any of the Stephen children to go into science, and not just because a scientific career was considered something below the elite. Science was a path for the middle classes rather than for those educated at Eton and Oxbridge. There were very few from the Eton–Oxbridge set who went into a salaried scientific career during this period. With the demise of the provincial amateur scientist, how would a passionate young entomologist, for instance, pursue their interest? Would they have been taken seriously in the academic world? Was science closing its mind to the enthusiastic gentleman and lady amateur? If Virginia, for instance, had wanted to devote her life to the study of insects, what would she have done? Where would she have studied? What careers might have been open to her? Would the same opportunities, or lack of them, have applied to her brother Thoby or to Lytton Strachey?

Lytton just stuck to reading *Nature* magazine every week and to thinking about history in a flexible and open way. The lack of opportunities in evolutionary biology meant that the subject was not attracting curious young people. In the past, the main profession for natural scientists, apart from teaching, was the church. To become a clergyman was out of the question for many. For aristocratic young men, there was not an easy path into the sciences as there used to be. For people such as Julian Huxley, the new institutes led to job opportunities, but Eton and Cambridge did not lead in that direction at all. Also, the force of the changes played a role in turning bright men like Fry and Strachey away from science and towards the arts.

On the other hand, having been brought up in a culture of observing and collecting insects and flowers, Virginia had developed a strong interest in natural history. One of her memories of her mother, who died when Virginia was thirteen, was of being taken to a public house in St Ives to buy a bottle of rum. On their return home, they used the rum to attract insects and drown them in a dish in the garden. They then pinned the dead animals on trays for identification and classification and thought about how flies and wasps might be related to one another.

While she was still a child, Virginia had been appointed secretary of the Stephen Family Entomological Society, chaired by her father. In this role, she had prepared the agenda and the minutes and learnt how to summarise the confused and confusing comments of the other members. She soon realised that identifying species and observing their behaviour was difficult and controversial and argued that it was important to record the changes and set them within the context of each particular discussion. At the society meetings, she was fascinated by how people at first listened to a subject and then spoke about it. She observed this and wrote accurate minutes of the meetings.

The work inspired Virginia to study entomology and to think about the ways in which people control one another and other species. During her time alone she observed the ants and wasps in the garden, looked closely at the rhythms of their interactions, how they gathered food and defended one another. Soon she was taking time to read about the habits and taxonomy of butterflies and moths, their life cycles and eating habits. She became confident enough to challenge the content of insect books that contained descriptive detail and systems of naming and classifying. Her writing about insects was to be much like it was about people: how they moved and interacted.

The youthful Virginia's view of insects reflected her own development as an artist and human being. She disliked how men of science objectified nature. Instead, she wrote of the insects' 'organs, orifices, excrement; they do, most emphatically, copulate'.[6] She wrote of moths and sparrows from their inside, not knowing or caring where her descriptions might lead. She wondered whether it might be safer for the inexperienced new residents of Gordon Square to stay with what they knew and leave the risks of dabbling in science to others. It was advice that could have come from one of the new neighbours, Ottoline Morrell.

Ottoline was a flamboyant socialite with red hair and a big personality who brought together many Bloomsbury intellectuals, both artists and scientists, when she moved to a house in Bedford Square in 1906. Roger Fry had first met her in Paris in 1904, and she supported him through difficult times, persuading him to use his scientific expertise to give a much-needed injection of vitality to the world of art. He told her that art in France had assumed new methods that involved

Figure 8.2   Ottoline Morrell (1902)

measurement and observation down to the smallest parts, looking to find the atoms.

With Ottoline's help, Fry realised that post-impressionism was an art form that used scientific discourse to enable new forms of expression. It emphasised colour, texture and substance and encouraged experimentation with feelings. Fry could see that the excitement embodied in post-impressionist art spilled over into music, ballet and poetry, and later into the novel. This excitement, he realised, came from science. Modernist high culture – radical, ground-breaking art – was being informed by science.

Fry fell in love with Ottoline, but the sentiment was not reciprocated as she was already in love with Bertrand Russell. By 1910, the first volume of Russell and Whitehead's *Principia Mathematica* was taking shape. That was also the year Russell was appointed as a university lecturer at Cambridge. Ottoline's husband was Philip Morrell, a Liberal MP and one of the few who opposed the forthcoming war with Germany. Ottoline and Philip entertained a great deal at the Bedford Square house and indulged in political and sexual behaviour that was more progressive than most parts of the enlightened London society. They became close friends with the Stephen sisters and, being about ten years older, guided the residents of 46 Gordon Square in modern living. They also introduced new friends such as D. H. Lawrence and Aldous and Julian Huxley.

***

'On or about December 10th 1910 human character changed.' Woolf explained why in her own inimitable way: 'In life one can see the change, if I may use a homely illustration, in the character of one's cook. The Victorian cook lived like a Leviathan in the lower depths, formidable, silent, obscure, inscrutable; the Georgian cook is a creature of sunshine and fresh air.'[7] More than that she didn't explain. Some historians have even suggested that the change was the new government in London, or even the accession of George V. It may be that the change was the opening on 8 November 1910 of Fry's post-impressionist exhibition at the Grafton Galleries. Whatever the cause or causes, it is sensible to say that the change was an historical shift, a confluence of cultures and interests, a zeitgeist.

It could be argued that major historical events occur randomly, caused by environmental or social change. These changes are often not recognised until some time after they take place. Whatever inspired Woolf's remark, her December 1910 event was much bigger than she could ever have realised at the time. A new middle class was emerging that savoured the tastes of highbrow and lowbrow alike. The high and the low had been mutually suspicious of the middle for generations. Now science and a little more equal opportunity were making possible stronger links between different classes, but for the Stephens and the Stracheys, the middle would always be a place of vulgarity and ostentation, a money-seeking, tasteless dystopia.

Meanwhile, another kind of split had emerged between the scientists who measured things and the others who were happy to describe them by whichever way they chose. The newly educated class tended to measure, and, as that professional group got bigger, some cautious people thought that society would not be able to accommodate so many technocrats. Vanessa Stephen, for example, warned that science was like sodomy: 'People are simply blindly prejudiced against it because they think it abnormal.'[8] As if to prove the point, in February 1910 Vanessa played a part in one of the most infamous hoaxes of the times, an escapade they saw as an experiment but which attracted a lot of amusement in the popular press and caused a lot of embarrassment. She and five others, including her sister and brother, blacked up, dressed as visiting Abyssinian diplomats and got on a train to Weymouth. There, they boarded the new warship *Dreadnought*, called on the captain and asked him to show them around. When the hoax was discovered it set off debates challenging many of the national values that people held very dear, the tradition of Empire, race, gender roles and national security. It opened up the ever-present split between control and liberation, between the establishment and culture, modernism and dissent. The group in Gordon Square were fearlessly challenging convention.

In November of that year, Fry's exhibition 'Monet and the Post-Impressionists' at the Grafton Gallery caused another stir. Although he had decided to drop science as a career, he had retained an interest and respect for its methods, especially its rigour in testing new ideas. The Cubists, who featured in the exhibition, were using new ways to record and share their observations, even to measure things like nature and human feelings. Their methods had moved on from the order and precision of photography and offered individualist styles with shocking detail. Familiar images were measured, analysed and dissected so that the bits could be reassembled to give many fresh perspectives of the originals. Unbeknown to any of these artists, at that same time, genes were showing up of the chromosomes in T. H. Morgan's New York laboratory, also ready for measurement, analysis and dissection before reassembling themselves, bit by bit.

Fry's exhibition asked the English middle class to replace their favourite picturesque flattery in art with a more challenging aesthetic. This was seen in some of the French impressionist pictures being shown together in London for the first time. Keynes's response to the exhibition was to be taken over by 'powerful and valuable springs of feeling' because it discarded many of the Victorian traditions that he was so eager to replace. Gone was chocolate-box idealism and Bentham, 'with his over-valuation of economic criteria'.[9] In their place were twenty-one Cezannes, thirty-seven Gauguins and twenty Van Goghs, all of which challenged the social order.

Many of the exhibition's critics thought it was a step too far. They couldn't understand where it was leading. Should artists be trusted to take such journeys into the unknown and report back to the public with their own strange experiences? Such untravelled pathways were likely to be dangerous for a society already stripped of its ethics and traditions. Realism was so much safer.

Several years later, Virginia Woolf would say, 'Let us record the atoms as they fall on the mind in the order in which they fall, let us trace the pattern, however disconnected and incoherent in appearance, which each sight or incident scores upon the consciousness.'[10] Science was influencing art and the novel.

Fry realised that form itself, colour and line, could be art. Indeed, form was the essence of art. Fry knew enough about biology to understand that, although nature was bounded by the morphology of animals and plants, every species and organism also needed a sense of place within its physiological, ecological and behavioural setting. With the exhibition, Fry used his scientific knowledge to show that the same anatomical structure can adapt to different functions according to changing influences such as the environment. The same idea was being explored in art as well as biology: structure was only part of a much larger world. The two approaches of art and biology were becoming part of a whole. The biology students were taught that 'structure needs function', a challenge to them over more than half a century. All the processes needed to be considered at once, a difficult task at the beginning of the twentieth century.

Many of the Grafton Gallery visitors were members of the upper middle class, still dependent on servants and income from invested capital. They were not prepared for the shock that the pictures transmitted. A critic wrote, 'One great lady asked to have her name taken from the Committee. One gentleman had to be taken out and walked up and down in the fresh air for five minutes. Fine ladies went into silvery trills of artificial laughter.'[11] *The Times* was more thoughtful: 'It professes to simplify and it throws away all that the long-developed skill of past artists had acquired and perpetuated. It begins all over again and stops where a child would stop.'[12] Fry was praised by those who wanted to see the world from a different perspective and criticised by those who did not. Surprised at the strong reaction to his efforts, Fry went out of his way to thank his friend Ottoline Morrell for her help in making the exhibition happen, though whether as an organiser or a muse was not clear.

By 1911, the social and sexual explorations of the wealthier Bloomsbury intellectuals had settled down. Virginia Stephen married Leonard Woolf the following year. Leonard had a maturing influence on the Gordon Square household. He helped Lytton Strachey realise that his career was to be in writing history. As a role model, Lytton took Virginia Stephen's father rather than his own and became a writer of historical biography rather than a scientist or a colonial civil servant. By then he must have been aware that the days of the Empire were coming to an end, as were the days of the gentleman scientist. He had not been offered a university fellowship, and his flamboyant lifestyle and expensive habits meant that he could not afford to live on the salary of a professional scientist in a research institute.

Roger Fry followed the success of the post-impressionist exhibition with a second show in 1912, this time with the first London presentation of Picasso and Matisse. He cheekily included some pictures by his friend Duncan Grant. At the same time, the Russian ballet was performing in London where a grand artistic reawakening was taking place. Fry was especially pleased to be able to talk about the artists and dancers in the same way as biologists. With new techniques and materials, new ways of collecting, displaying and analysing data and feelings, both artists and biologists were beginning to support one another and make connections between their fields.

In 1912, H. G. Wells, secretly angry about the success of the art exhibitions, persuaded Fry to contribute to a volume about *Socialism and the Great State*. Authors from different specialisms set out ideas of how their work should be used and financed. Both art and science needed investment before any gain, let alone critical success, and there was much discussion about how that was to be done. Wells painted a picture of what he saw as the direction in which their society was being taken by science, moving towards his own kind of socialism, what he called 'normal social life'. He saw this as a situation that occurs naturally, one that cannot be created under instruction. To discover new things or ideas is a special gift that is bestowed on artists and scientists alike.

Lankester and Fry would not budge from their belief in an individual's responsibility. Wells, meanwhile, clung to his more collective socialism. The question of whether humans are organised best as individuals or in groups was taken up by Karl Pearson and Raphael Weldon. Pearson and Weldon's work signalled the birth of behavioural genetics, which considered the question of altruism leading much later to the concept of the selfish gene. A fear then was that mass production from machines would increase materialism at an unprecedented rate. Some people were afraid that the big state machine should take over from the efficient handiwork of the individual. Was human happiness and pleasure going to be lost in a socialist world of the future?

So, while some scientists were slowly entering new and unusual areas of study, artists were being challenged to create new forms. By 1913, Fry was forty-seven years old and an experienced and talented public speaker. In a lecture at the Queen's Hall, he argued that art was at a major crossroads. Art, like western civilisation itself, was entering a period of crisis. This was Fry's message to the new middle class. He was excited about offering them a popular art that they could take into their homes. To do this, he advocated taking art into the commercial realm.

Convinced that the individual artist should sell his own artwork, Fry launched the Omega Workshop in Fitzroy Street in the spring of 1913. In the workshop, Fry designed and made prints, furniture and other objects. It was an exciting project that brought together artists from different backgrounds. They got on well, at first.

The idea of the workshop originated from the English Art and Craft movement. This had become an important agency in the lives of young artists in southern England because it excelled in design and printing, albeit on a small scale. The Fitzroy Street workshop involved several local artists, including Wyndham Lewis, Vanessa Bell and their friends, all of whom were confident that their work brought joy to those who bought it. When Fry was asked to explain this, he gave a very biological explanation: art brought out the human in people, what was left over once they had earned their food and warmth and had sex. As a former Apostle, Fry believed that art was the disinterested love of truth and it required honesty.

The honesty of these Omega Workshop artists was soon to be tested by an incident that exposed the different ethical and social values of the Bloomsbury group. It began in the autumn of 1913 as an apparently trivial argument over the creative rights to be gained from one of the designs at the workshop but subsequently became an important sign of deep divisions in social attitudes. The workshop had been commissioned to decorate a whole room at the *Daily Mail* Ideal Home Exhibition. Wyndham Lewis and his friends thought the work was for them. Fry thought it was for his other set of friends. It was a common confusion, which could have been resolved by a simple division of labour. But not so this time: *'C'est trop fort!'* shouted Fry, slamming the door in Lewis's face. Lewis and Spencer Gore walked out, deserting the Bloomsbury artists forever and moving up the road to Camden Town. Lewis's group was much more socially aware than Fry's and took commercialism very seriously, which explained the move away from the highbrow residents of Fitzroy Square.

Similar conflicts were arising in other parts of the cultural scene. Another friend of Lytton Strachey was E. M. Forster. Forster's 1910 book *Howard's End* portrayed an ambitious young office clerk, Leonard Bast, who threatened the stability of the Schlegels, two sisters and a brother. Together, they resembled the Stephen siblings. The book was written in a realist style, with lots of facts and details, something that Virginia Woolf disliked because it kept 'life' out of the writing. Without life, she said, nothing was worthwhile. Woolf wanted to shine a light on the dark places of psychology, character, thought, desire and memory.

A similar criticism was often made about H. G. Wells: a lack of human insight. This was not quite fair. With *Boon*, Wells offered a bitter satire on social convention. He called the novel his 'age of acquiescence', but others were outraged by its arrogance and disrespect for the previous generation and its fledgling middle-class artists, particularly its direct attack on his erstwhile friend Henry James. When the *Times Literary Supplement* asked him to review the book, James used the opportunity to retaliate with scorn and derision. His review ended with his public announcement that he was ending their friendship. In response, Wells boasted that, 'I bothered him and he bothered me.' Wells contended that James opposed materialist artistry because it had 'so much life, so little living'.[13] In turn, James joined Woolf in accusing Wells of not accepting art and beauty as ends in themselves.

Wells did not belong anywhere. He had left his class, both physically and in spirit, and he was now lonely and sad. He mistrusted his new acquaintances and even despised some of them. Often his feelings for his contemporaries were divided between admiration and envy, longing and scorn. Those on the other more privileged side of the fence had different reactions to intruders like Wells. Woolf felt that they were trying to seize as many opportunities as they could. Woolf accused Wells of being interested merely in:

> the fabric of things who has given us a house in the hope that we may be able to deduce the human beings who lived there.[14]

The influential critic Edmund Gosse also felt that Wells had cut loose from literature. But Gosse had done well at cutting himself loose too, in 1907 having written *Father and Son*, a biography of his cruel father (a marine biologist who had described several new species of sea animals).

Similar changes were taking place in the lives of some of the other Bloomsbury artists with humble backgrounds. Wyndham Lewis and Jacob Epstein set up the Rebel Arts Centre at 38 Great Ormond Street. Women played an important role in the centre, although the painters Kate Lechmere and Helen Saunders had to work hard to challenge the common association of creativity and masculinity. They campaigned for the vote and explored such topics as sexology and eugenics. Their discussions were reported by newspapers, adding to the public's curiosity about their strange new work. The *Daily Mirror* published a photograph of Lewis at work with the caption: 'With these revolutionary works it is not always possible to tell "t'other from which" and until the average man can learn to penetrate their meaning he will probably pin his faith on the old schools.'[15] The Rebel Arts Centre was given an important lift to its morale by a young Frenchman who had come to Bloomsbury in 1911: the twenty-year-old sculptor Henri Gaudier-Brzeska. He admired the primitive statues at the British Museum and wanted to bring temporality, sexual aggression and dynamic animal competition into his art. Ancient sculpture was radical and highly expressive of the time. Ancient sculptors worked from solid lumps of hard rock, 'direct carving' as they called it, making virile objects that shocked and offended some tastes. Influenced by this, Gaudier produced aggressive metal objects with names such as *Knuckledusters* and *Doorknocker*. He integrated biological observations of behaviour and sexuality into sculpture in works like *Duck*, *Cock*, *Stags*, *Bird Swallowing Fish* and *Torpedo Fish*. A thirty-one-year-old American sculptor, Jacob Epstein, meanwhile, made a weird prediction of war in his *Rock Drill*. *Rock Drill* was an expression of what humans had become in 1911, and was inspired less by biology than by engineering. The work looked more like an early machine gun than a rock drill, curiously about four years ahead of its time.

Together with some other young French artists, Gaudier had studied biology, was strongly influenced by Henri Bergson and expressed his own energy through a confidence, even a sense of superiority. He epitomised the angry young rebel with plenty of understanding but not much feeling. There was always an inner tension between him and Epstein at the Rebel Arts Centre, on one hand, and Fry's friends at the Fitzroy Street workshop, on the other. These artists began marketing their art objects in direct competition with one another. Lewis famously rebuked the Omega Group as a party of strayed and dissenting aesthetes.

In 1911, Henri Bergson had given a lecture in Bloomsbury about his book *Creative Evolution*, published earlier that year. Favouring some of the implications of Lamarck's linear speciation, Bergson thought Darwin's adaptation by selection evaded evidence of creativity in early humans. He advocated the idea of *l'élan vital*, a creative force driving evolution, an idea popular in France but little known in England. Bergson argued that life was a continuous surge of energy, made up of free spirits that underlay nature, not a set of rules leading to a fixed end. Matter was at the bottom of his list of priorities and spiritual emancipation at the top. He thought that living systems rose up this slope.

In 1912, Wyndham Lewis adapted Bergson's ideas by fusing time and space to make images of geometric order within chaos. This was the time of vorticism in art, where drab colours were whipped into brain-whirling orgies of confusion. In this art, time had an unseen role. Like other neo-Lamarckians, Bergson thought the origin of an organ such as the eye would not be explained by physics or chemistry but instead by psychology. That was the philosopher's view, not attempting to analyse or test by experiments.

One of the first students of Bergson's philosophy was Karin Costello. Costello was born in 1889 in Philadelphia, and at twenty-one she earned a first in moral sciences at Cambridge. She was encouraged to stay on in England by her uncle Bertrand Russell, and she accepted a fellowship at Newnham College. There, she studied Bergson's philosophy, search-ing for patterns in the complexity of self-organised systems. She thought pattern was something that may even have been implicit in the work of Russell and his mathematics.

Costello moved into a flat in Great Ormond Street. In 1913, she was introduced to Virginia Woolf's younger brother Adrian Stephen. They started to see more of one another and soon developed a strong intellec-tual rapport. They talked incessantly about Bergson's *Creative Evolution*, particularly the idea that *l'élan vital* might be what Adrian called the 'psy-chological cause'. It was a popular idea, one that helped many young peo-ple without a religious belief to give some meaning to life. Most observers saw these ideas as incompatible with Darwin's scientific theories.

However, in August 1914, the people of Europe were overshad-owed by events beyond their expectations, let alone their control. Like Karin and Adrian, so many were looking forward to a phase of intellec-tual breakthroughs and economic growth, using science and technology to spread more comfortable styles of living to more communities. They shared an interest in psychology, and they wanted to make a mark for themselves, something they could accomplish together away from the traditions and constraints of their families. Many ambitions such as theirs were being nurtured with a spirit of fresh optimism. In October 1914, Karen and Adrian were married and eventually became famous for their work together. But events ensured that their interest would be put to an unexpected use. Their optimism was suddenly dashed and twentieth-century world history forced a tragic course for millions of their contemporaries.

9
Time passes, 1914–18

The war made a slow start for H. G. Wells and Roger Fry. They could only sit and be fearful. Wells's 1914 novel *The World Set Free* opens thus: 'The history of mankind is the history of the attainment of external power.'[1] The novel goes on to imagine what would happen as a result of the latest discoveries about the internal energy of the atom and to express the terror arising when bombs were dropped from aeroplanes. It was inevitable in such times that Fry's Omega project struggled to survive.

One of the first bombs of the First World War dropped on Bloomsbury, at the Dolphin pub in Lamb's Conduit Passage. The clock above the bar stopped at 10.40, and several people were killed. From then until the war's end, Zeppelin airships brought fear to the streets. Wells escaped London to a house on the Essex coast with his new lover, Rebecca West, only to find that they were living directly under the flight path of the airships on their way to bomb London. He stood on the balcony of their house, looking east and dreaming of the new utopia that many predicted would follow the war.

Lankester claimed that the war knocked him off his perch, for it brought an end to so much that he held dear: regular friendships, individual freedom and scientific contributions to human progress. More practically, it severely reduced the value of his pension and put an end to his weekly column in the *Daily Telegraph*. It made him feel lonely and depressed, or as he said, 'It has laid me flat.'[2] With many of his old friends and colleagues busy with the conflict, it was hard for Lankester to know what to do with his time. Although there was still the *Quarterly Journal of the Microscopical Society* to edit, only a few manuscripts were being submitted, and he was getting a bit too old to command the full respect of the Society. Consequently, he spent more time with the Committee for the Neglect of Science, which helped get his 'Easy Chair' articles published as short books that soon became best-sellers despite the paper shortage. Just one of the many collected editions of his articles was reprinted four times during the war, but he continued to feel angry about losing his job at the museum and he found it difficult to settle into a quieter lifestyle. He must have been afraid of his own excesses.

As the war went on, Lankester became more and more anxious about its course. The air raids made it difficult for him to stay in London. First he stayed with Jane Wells, but the difficulty she was having with her husband made it inappropriate to remain for long, so he moved down to the Dorset coast. At least, there, Ray had a chance to collect specimens of marine animals from the tidal flats and to talk about them with local specialists, but he wasn't suited to provincial life and easily tired of the locals: 'Bournemouth is the dullest place in the British Isles. Rank respectability, pasty-faced and expressionless people roll along here in countless numbers and never smile. They are hideous and make one feel nausea.'[3] The only project that Lankester completed during the war years was his work with prehistoric flints at Henry Stopes' archaeological site in Kent, but the flints were inert objects and failed to excite him in the way that living creatures did. He continued to publicise as best he could the big ideas of evolution by natural selection, giving public talks and keeping an interest in the work of the British Museum. It was he who was responsible for keeping it open throughout the war. The prime minister, Herbert Asquith, wanted the museum closed but accepted the argument from Lankester's lobby to keep it open.

In 1914, Karl Pearson took the new chair of eugenics at UCL that had been endowed by Francis Galton. Until the outbreak of the war Pearson had been preoccupied with the possibility that a higher state of civilisation could come about through a struggle between one race and another, leading to the survival of a physically and mentally fitter group, but he was now too old to continue with these projects and withdrew from political and scientific work to concentrate on writing his biography of Galton, volumes which he funded himself and published in 1914, 1924 and 1930.

***

Many young artists and scientists in Europe rose up to fight with enthusiasm. Henri Gaudier-Brezska left his studio in Great Ormond Street to join the French army, and, within a few weeks, he was writing from the front line for the Christmas number of *Blast* magazine:

> Just as this hill, where the Germans are solidly entrenched, gives me a nasty feeling solely because its gentle slopes are broken up by earth-works, which throw long shadows at sunset ... I shall derive my emotion solely from the arrangement of surfaces.[4]

He died in one of those trenches at Neuville-St-Vaast in June 1915.

The war stimulated interest in the practical application of chemistry and physics to methods of warfare. There were fewer obvious applications of biology apart from medicine and nursing. Some doctors and nurses were able to offer psychological help to soldiers, but there was so little knowledge of conditions such as shell shock that there was not much to be done other than give love and care. Often the patients they cured were sent back to fight. The unlucky ones might be shot for cowardice.

Many of the young generation that went to the front did not share the general melancholy. The war was an altogether different matter for one new graduate from Oxford, J. B. S. Haldane, known as Jack, who became a famous Bloomsbury scientist. He had studied maths and classics, was part of an aristocratic family of Scottish baronets, and had been brought up to relish fear. He enjoyed showing off his fearlessness to his uncle who was minister of war. He looked forward to using the real battlefield to try out some of the physiology experiments he had done with his father, J. S. Haldane, a professor of physiology at Oxford. He joined up as an officer with the Black Watch as soon as he graduated and went off to the front in Belgium, taking equipment to monitor respiration and other bodily functions affected by poisonous gas. Before the battle of Aubers Ridge in 1915, Haldane, or Rajah the Bomb, as he was known by his men, wrote to his father: 'I am enjoying life here very much. I have got a most ripping job as a bomb officer.'[5] Haldane said he had a good war and raided the German lines at night by throwing grenades into their trenches. The war seemed to make little difference to the way he lived and thought; working with dangerous chemicals was just the kind of thing that he was used to and only the accommodation was different. What most people thought to be uncomfortable and frightening did not seem to bother him. He saw the war as a controlled experiment in which the guinea pigs were men.

Unable to go to the front as a voluntary nurse, Jack Haldane's sister, Naomi, worked at St Thomas's Hospital. When she was eleven years old, in 1908, Naomi and her brother began breeding real guinea pigs in their Oxford garden. They spent a lot of thought monitoring the inheritance of particular features as a serious scientific experiment. It turned out to be some of the first work ever done on the genetics of gene linkage and crossing over, and it confirmed the concept of genes as measureable hereditary particles. They were writing up the results when Jack went off to the war and Naomi moved to London. Wanting to continue the experiments, she took some of the guinea pigs with her and kept them in hutches on the river terrace just opposite the Houses of Parliament. It was a very significant breakthrough, completely overshadowed by the war, and was never properly acknowledged.

Also threatened with obscurity were the young men of those years who wanted to fight but were prevented by bad health. One such was Ronald Fisher, the bright young Cambridge mathematician who was friendly with Leonard Darwin. He had left university in the summer of 1914 to work as a statistician in the City of London. Even his good record as a part-time officer in the Territorial Army could not prove to the Recruitment Board that his eyesight was good enough. Reluctantly he settled for teaching physics and mathematics to cadets. He rented a cottage out in the country with his seventeen-year-old wife Eileen, and also helped the war effort by farming. Living in the country gave him the opportunity to fulfil his own political ideals: if you believed strongly in eugenics, and if you and your partner were healthy and intelligent, you had a duty to society to have many children. Propagating might help make up for the high number of officers being killed, the other ranks being less of a concern.

The quiet country evenings allowed the Fishers to concentrate on their work and raise children. There, Eileen read to him at breakfast from *The Times*, worked as his typist, housekeeper and laboratory technician, cleaning out his experimental animals' hutches. She had found her hero in Ronald and soon he went out to teach biology at Bradfield College, nearby. Eileen stayed at home with her sister to look after the pig and the cow. In the evenings, they read classical Greek, and Roman and Scandinavian literature, while all three kept fit by playing with a medicine ball every day before breakfast.

Through these war years, Fisher continued to try and solve the problem about heredity that had kept the scientists of Cambridge and London apart. The source of the problem, he recognised, was the difference between evidence for inheritance from single characters of mutating genes and evidence from questionnaires about things like height and intelligence. He wrote an article attempting to reconcile the differences between the Cambridge geneticists and the London statisticians, focusing on the use of new methods to analyse the data.

However, the plan backfired, only to proceed in an unexpected way. Fisher's article was sent to be refereed by Bateson's colleague Reginald Punnett in Cambridge and by Karl Pearson in London. The two enemies recommended separately that it should not be published. The angry author bemoaned a suspicion that 'the rejection of my paper was the only point in the two long lives on which they were ever heartily at one'.[6] It did get Punnett back on speaking terms with Pearson, and in 1916 Fisher sent his manuscript to be published in another journal. This was run by another group of young men who were focused on targets and who wanted measured results. To them it was sensible that mathematics should be involved in any scientific problem.

***

At the beginning of the war, the Bloomsbury ecologist Arthur Tansley was forty-three years old and worked at the Ministry of Munitions. He tried to keep tabs on the struggle between plants and their changing environment, and he continued to monitor his long-term experiments near the Cambridge fenland. He also went up to the Norfolk coast and continued his surveys of woodlands and heath in other parts of Britain. He was doing something more than making static descriptions of the morphology and compiling lists of species. He was also comparing these data to the physiological, pathological and genetic features of the organisms he studied, integrating as much as he could from all these disciplines.

At times, the struggle of daily life depressed him. Worse, as the war went on, it was becoming more difficult to escape to the countryside. He was traumatised by what had happened to so many young men in the war and he began to have restless nights haunted by vivid dreams. One of these influenced him so deeply that his mind became deeply troubled.

> I dreamed that I was in a sub-tropical country, separated from my friends, standing alone in a small shack or shed which was open on one side so that I looked out on a wide-open space surrounded by bush or scrub. In the edge of the bush I could see a number of savages armed with spears and the long pointed shields used by some South African native tribes. They occupied the whole extent of the bush-edge abutting on the open space, but they showed no sign of active hostility. I myself had a loaded rifle, but realized that I was quite unable to escape in face of the number of armed savages who blocked the way.

Then my wife appeared in the open space, dressed entirely in white, and advanced towards me quite unhindered by the savages, of whom she seemed unaware. Before she reached me the dream, which up to then had been singularly clear and vivid, became confused, and though there was some suggestion that I fired the rifle, but with no knowledge of who or what I fired at, I awoke.[7]

This dream encouraged Tansley to read the new medical journals and identify the most important work being done in the emerging field of psychology. On the basis of this research, he wrote about the effects of war on the human mind in *New Psychology and Its Relation to Life*, and it was published in 1920. Wounded soldiers suffered great psychological distress, a fact that was usually not recognised or understood. There was plenty of public interest in possible explanations for soldiers' mental trauma, which made Tansley's book a best-seller and one of the standard introductions to the subject. The book's success caused Tansley to draw comparisons between psychology and ecology, but he soon became aware that the former was attracting much more attention and praise than the latter. Only a few specialist scientists read his ecological work with as much enthusiasm. Their reactions to this contributed to his turning attention towards psychology. The very implausibility of Tansley's involvement in psychology made him representative.

Tansley was intrigued by the new theories in psychopathology. By his own account, however, his knowledge owed more to personal experience than study or research. In 1916, aged forty-five, married with three daughters, secure in his profession and having recently attained as good a reputation as any scientist could wish for, he had that haunting dream. It impressed him very deeply and led to a resolve to read Freud's work, including Freud's sexual theories. Tansley was looking for his next move.

The conflict in Tansley's mind between ecology and psychology became more difficult with the news that he had been elected a fellow of the Royal Society for his work in Norfolk. The new confidence this gave led him to follow Lankester's example of twenty-five years earlier and reform the teaching of biology at Cambridge by introducing experiment and analysis of results, as well as more physiology and cellular work. These reforms did not go down well with traditional observers who saw any process that needed timing, weighing or measuring as threatening. The 1917 'encyclical' in *The New Phytologist*, signed by Arthur Tansley and Francis Oliver, pleaded for a vitalised and practical curriculum based on plant physiology and ecology alongside morphology rather than subordinate to it.

Tansley's reforms were denounced by some as 'Botanical Bolshevism'. His zeal was a significant factor in his not being elected to the Sherardian Chair of Botany at Oxford, for which he was a candidate in the autumn of 1918, a professional setback that he later regretted: 'I've been getting some experience in the "Gentle art of making enemies" lately … Reactionary forces are pretty strong here, and it will be a hard struggle to get anything progressive done. But I am going to have a good try.'[8] Even with war, the old male establishment did not admit that its values were under threat or that they would have to change. People like Tansley were trying hard to change the academic system, yet their Oxbridge counterparts were resisting.

***

Julian Huxley was a much more relaxed and rounded character than most of his wartime contemporaries. Following in the footsteps of his Ray Lankester, he studied zoology at Oxford and then stayed on to research the migration patterns of water birds. In 1910, when Julian was twenty-three years old, he began a series of experiments monitoring the great crested grebe. His careful observations showed that the bird's mating behaviour had been shaped by evolution: the male and female, having similar plumage, had evolved to perform an elaborate dance. The resulting publication in 1914 became a classic in the genetics of animal behaviour.

With his pedigree and background, he was soon invited to set up the first biology department at Rice University in Texas. When he arrived in America in 1912, Huxley went to Harvard to meet Sewall Wright, a modest mathematician looking for evolutionary trends in guinea-pig data. Looking together at some baffling evidence, they proposed that certain characteristics might be controlled by genes on the same chromosome. The concept of genes as particles that coded for biological chemicals such as enzymes, in turn leading to structural characters, offered a workable approach to this work on families of experimental guinea pigs. The young Huxley knew, however, that to find evidence for such genetic control would be difficult. Wright, too, was cautious. He delayed an announcement of their ideas until they were more sure.

Huxley's work in Texas meant that he did not join the war at first. The call to arms was strong, however, and he returned to join the intelligence corps. During his training, he wrote of being pleased to feel physically fit, and then the reality took over: 'In the spring we were sent to a camp at Upstreet, near Canterbury. I remember riding about the peaceful Kentish lanes, lined with white May bushes and pink-flowering horse-chestnuts, in strange contrast to the distant boom of heavy artillery from across the Channel.'[9] In general, the effect of the war on the Bloomsbury scientists was to focus minds on work and to seek clearer and more objective targets. The pain of their friends' deaths was intense, as was their own guilt and loneliness as survivors. Haldane, the toughest of the group before the war, retained his need for action and his eccentric determination. Fisher became even more committed to eugenics, which helped him build even closer links to fascism. They were all hardened by the war and developed strong political aims that were to influence their work for the rest of their lives. The scientists were not alone in being forced to re-examine their place in society as a result of the war. Bloomsbury artists and writers struggled even more to find a social role and a future.

Full of the catastrophe of the war, a group of Bloomsbury artists produced the magazine, *Blast*. The first edition was published in July 1914, two months before the war began. *Blast* was edited by Wyndham Lewis, who wanted it to be seen as a reference for English creativity. The magazine celebrated the new age of mechanised printing by including illustrations, wood-cut prints and other kinds of graphic illustration, but the war meant that *Blast* only lasted for one more edition, a so-called 'War Number' published in July 1915. The project had hit its target, the young artists had identified their enemy to be the bureaucrats and their war was also what killed it.

A month after this final publication of *Blast*, Philip Morrell gave one of the bravest speeches ever made to the Westminster Parliament. In his speech, Morrell set out the case against Britain joining France in war with Germany. Ottoline wrote in her diary, 'I can never forget seeing him standing alone, with nearly all the House against him, shouting at him to "Sit down!"' He was one of the few establishment figures courageous enough to speak out publicly against his country's decision to go to war.

Some young men who could not face the war were welcomed as the Morrells' guests to enjoy thoughtful weekends at Garsington, their Elizabethan country house near Oxford. Ottoline Morrell's French teacher, Juliette, described the house:

> Six miles from Oxford, on top of a steep hill, stood the little village of Garsington, with its old Manor House. This was an unspoilt Elizabethan stone house with three gables, approached by a court flanked by two immense yew hedges. Bought by Philip and Ottoline Morrell in 1915, the house was woken from its Tudor sleep and became alive, personal and loved. The dark Tudor panelling of the large sitting-room was painted a glowing Chinese red and the narrow grooves gilt; the tall gothic windows were framed with yellow

and flame-coloured curtains, the floors covered with gold and salmon-pink Samarakand rugs. Chinese lacquer cabinets created a symmetry of black and gold, and logs burnt in the stone fireplaces.[10]

In the same piece, she went on to describe Julian Huxley, soon to become her husband, and his brother Aldous: 'I was amazed at their difference – Julian ebullient, forthcoming, putting himself out to entertain; Aldous reticent, gentle, often remote, but both with innate gifts of high-powered intellect and imagination.'[11]

Aldous Huxley was born in 1894, went to Eton, and nearly became blind when he was sixteen. After Oxford, he was invited by Ottoline Morrell to Garsington, and there he met D. H. Lawrence, whom he knew to be a genius, championing individuals from his own working-class background. Science and objectivity were of no importance to Lawrence, and instead he stressed the love between individuals and kindness within families. Julian and Aldous were not so sure: their brother Trevelyan had committed suicide in 1914 after a period of depression. With setbacks like that, and the war, they were pleased to be in a family of peaceful people.

During the war, science was enabling production to be central-ised, and this undermined individualism in farming, industry and the army. This fitted the new ideology of socialism where camaraderie was forging groups of like-minded young people, what some thought were comparable to family groups. The writers, Lawrence and Aldous Huxley, however, were wary of each other. Lawrence was concerned about how people felt about themselves and one another; Huxley remained distant from those he knew and from the characters he wrote about.

When Aldous Huxley fought with others, he did so bitterly, and his writing upset some of his subjects. His satirical novel *Crome Yellow*, pub-lished in 1921, shared some of the dilemmas posed by the clashes of sci-ence, social class and war: 'Simultaneously the same person is a mass of atoms, a mind, an object with a shape that can be painted, a cog in the economic machine, a voter, a lover.'[12] Lawrence was irritated by this kind of endless speculation, arguing that it served no purpose and raised false anxiety. It was the same argument he told in his earlier 1913 book, *Sons and Lovers*: 'Science makes sex something to be serious about' though 'amour is more fun.'[13]

Aldous clearly enjoyed taking swipes at those who clung to Victorian habits of mind. His next book, *Antic Hay*, published in 1923, savaged many of the people he had known at Garsington and other members of the chattering classes. He was one of the first of the Bloomsbury group to notice that the new scientists were looking to the intelligentsia for recognition, preparing to be accepted by them as essential driving forces in the new world. Wells was doing the same thing, but he was more of an outsider and appeared to be thick-skinned enough not to care what people thought of him. Huxley was more sensitive and looked towards the new civilisation in California for escape from the society he criticised.

Aldous did not get on with H. G. Wells. In his 1915 novel *Boon*, Wells made a cheeky reference to the Garsington group, especially those who were Cambridge Apostles. He said their philosopher-leader, George Moore, had 'played uncle to so many movements and had been so uniformly disappointed with his nephews'.[14] This was the most powerful group at Garsington: bright, eager and creative intellects not believing in war but used to protection from a hard life. Most of those who visited Garsington were strongly committed to pacifism, especially Bertrand Russell, James Strachey and his brother Lytton. One day all three met and fell into one another's arms as co-workers in a great cause. With his confidence wounded, Russell said then that he thought all philosophers were failures. At least he was ready to admit it. More than that, Russell said his career was a threefold failure: he abandoned religion and objective ethical knowledge, he accepted that mathematical knowledge was tautological, and his defence of scientific knowledge was limited.

While at Garsington, Russell decided to leave mathematics. He wanted to be more active in promoting his views about pacifism and how it involved eugenics. Being a conscientious objector convinced him that humanity was on the road to the socialist utopia. Some compared this to the belligerence of those Bloomsbury artists known as the Vorticists, suggesting that its popularity would encourage an exodus of other young men from science. It was for more materialist reasons that Lytton Strachey had left his early interest in science for history and Fry had left botany for art. Meanwhile, H. G. Wells left zoology for failing his exams and Aldous Huxley left biology with bad eyesight.

There was some relief for the few who remained in the scientific world, especially now that its scope had widened to include psychology. The war was to inflict sufficient damage on people's minds to open new treatments for those fortunate enough to have access to the limited psychological services on offer. From the Bloomsbury group, Adrian Stephen trained to be a psychiatrist, James Strachey a psychoanalyst, and Arthur Tansley hovered between applied psychology and plant ecology.

Julian and Aldous Huxley appeared to be well-rounded people like their grandfather, one becoming a successful scientist in his own right and the other a successful writer. They were an important example of the concept of 'the two cultures' during the 1910s and 1920s. Although they had reacted differently to the death of their middle brother, Trevelyan, they were both interested to know how those strong reactions could be explained. They knew how important their grandfather had been as a scientist and how he dealt with his own periods of depression. Many of the characters in Aldous's novels were scientists who discussed how that made them feel. Similarly, Julian's holistic approach to evolutionary biology was as much a masterful stroke of the artist's brush as it was an interpretation from all the scientific evidence.

The war forced many people to think about what it meant to be human. In this conception, art and science worked together and could be understood as part of a whole system. James Strachey's best friend and fellow Apostle, Rupert Brooke, talked of a dialectic between the real and the ideal. Objectively, the Haldanes researched theories about genes, Tansley tested theories about ecosystems, and Julian Huxley looked for a grand law encompassing all of these pieces of information into one modern synthesis. Ray Lankester had done the same with his grand idea of degeneration, and, before that, Charles Darwin had given the idea of a single tree of life. Virginia Woolf said it was truth that war left behind in an individual, after feeling and thought were taken away. Like war, evolution was a game to these scientists and artists alike, using particular rules that kept changing and developing and in which the winner was the relentless opportunist.

10
The one culture, 1920s and 1930s

Many Bloomsbury intellectuals who survived the First World War sought to find a way of re-establishing networks between artists and scientists in order to connect with the past and to build a new future. There was a strong feeling of respect among many writers and scientists, and they saw the common patterns in their work. There was an increasing interest in listening to new groups involved in biology, and this brought together specialists best known for their very different views. J. B. S. Haldane and Ronald Fisher, for example, now found much common ground. There were also many experienced professionals positively encouraging new blood to rejuvenate those disciplines damaged by the war. Even the ageing Ray Lankester inspired more than his fair share of young followers, including H. G. Wells's son Gip. As though to seal a new sense of unity within the biological sciences, Julian Huxley started to bring together all the disciplinary ingredients for his 1942 classic *Evolution: The Modern Synthesis*.

In Bloomsbury, this surge in scientific interest and activity involved many more professional biologists and writers than before. These people were even more determined to explore the meanings and values of life, with an emphasis on molecular science and psychology. Many had developed new interests in Darwin's theories of natural selection in their search for the natural laws that underpinned life on earth, but positive evidence for natural selection remained elusive. Articles appeared in the press asking questions like 'Is Darwinism at Dusk or Dawn?' The paucity of direct evidence to prove natural selection drove more observers to reconsider Lamarck's ideas about acquired characters that he had set out in 1809.

Through the 1920s, there was a rise in interest from those looking for smaller and smaller clues. Many of those latest researchers were looking to biochemistry for evidence from psychology and genetics, so when the secretary of London Zoo, Peter Chalmers Mitchell, asked Lankester along to a meeting of the Aristotelian Society, they were pleased to see its members looking outwards at the grander picture. One after another, the speakers applied Darwin's ideas about evolution to social matters such as economics and politics. If animals and plants adapted to changing environments so could humans, even if it was so as to live in austerity and to die in wars. The proceedings of the meeting are still available as a book called *Life and Finite Individuality*, with the arguments for and against whether physical, biological or psychological categories are reducible to molecular explanations. Chalmers Mitchell saw a clear trend to a synthesis, as did another morphologist, the eccentric Scottish biologist d'Arcy Thompson. However, J. B. S. Haldane's father, the Oxford physiologist J. S. Haldane, thought differently. Thus, the Oxford tradition of reductionism was founded, and it continues now with Richard Dawkins' popular books.

George Bernard Shaw continued to be repelled by the apparent randomness and purposelessness of Darwinian natural selection and remained an advocate of Lamarckian evolution. His five plays *Back to Methuselah,* published in 1921, expressed his distinctive mood of despair at science and war. In his long preface to the plays, Shaw argued that humanity was heading towards catastrophe and explained his dismay that scientists thought 'the world could make itself without design, purpose, skill or intelligence'.[1] In the plays that followed, Shaw argued that evolutionary change occurred because it was needed, or wanted, just as Lamarck had suggested 100 years before. He still dismissed Darwin's theory, and Weismann's more recent demonstration that genetic information needed sexual reproduction to pass from one generation to another. It followed that he also ridiculed those who dismissed the concept of a utopia.

Much to Shaw's dismay, most Bloomsbury intellectuals, including Lankester, Fry and the Huxleys, accepted the argument that inheritance worked by the transfer of information from the genetic material to protein, not the reverse. The latter was the domain of Shaw, who persisted with eccentric support for the untestable, on the basis of what he called old-fashioned common sense. This made him many enemies, and the Bloomsbury insiders kept him at a distance, especially at informal meetings in their homes. That was where he liked to provoke people with conversation that roamed across disciplinary boundaries between the arts, the sciences and, of course, politics.

Aware that London had no social refuge for their well-connected socialist friends, Leonard and Virginia Woolf agreed to start a new club where young men and women could meet and talk about the arts and sciences. The first members included the archaeologist V. Gordon Childe and the writers Rose Macaulay and Aldous Huxley. It was called the 1917 Club and was at 4 Gerrard Street in Soho, close to Bloomsbury and even more bohemian. Through the early 1920s, many well-known Fabians joined the conversations: Wells, Ramsay MacDonald and even Shaw. James and Alix Strachey were also frequent visitors. Unusually for those days, there were about as many women as men, and science was represented almost as strongly as the arts.

The Woolfs also liked Soho, which was only a ten-minute walk from Gordon and Fitzroy Squares, where Keynes, Lytton Strachey and Virginia's sister still lived. These people were all too aware that the war had taken so much of what was good in life, leaving many in grief and despair. They believed that everyone now had to turn their attention to new growth. In 1919, Lytton Strachey feared that 'the whole square will become a sort of college';[2] the young survivors were realising only slowly that chemistry and physiology were needed for growth, that food was required for organisms to digest and come back to rebuilding life. The war had put stress on many equally necessary rhythms of daily life. It took some time for necessary routines to be re-established and settle down. Many in the group were drawn to the consoling elements of G. E. Moore's philosophy, especially Keynes and E. M. Forster, who continued to be defensive of Moore's values, particularly his candour, humour and rigour.

The Bloomsbury artists, with their Cambridge origins and their comfortable outlook on life, were distinct from the middle-class Fabians. They inspired other Apostles who settled in Bloomsbury after the war to set up another group, more elite than the membership of the 1917 Club. This second club was established in 1920 and came to be called the Memoir Club. It met only twice a year, and the members dined at one another's homes. Appointed speakers read from notes telling stories of their relationships, ideologies and attitudes to other people. In keeping with their rule of confidentiality, most records of the presentations and discussions were destroyed, but some notes survive, for instance, Keynes's frank accounts of his role at the Versailles peace talks in 1919.

Science was regularly on the agenda at the Memoir Club. One evening, the club met at Roger Fry's flat opposite Russell Square Station. Virginia Woolf's diary tells how Fry offered them 'overdone and tough duck'.[3] Then he told them to sit on the floor between the stacks of unhung pictures. Scattered all over the floor and tables were brushes and unfinished paintings, ink bottles and manuscripts, dirty coffee cups and saucers. They discussed the relationship of mystery to science. Virginia spoke for modern science, accepting 'the complete relativity of everything to human nature and the difficulty so many people had of talking at all about things in themselves', to which Fry argued that 'science can only begin when you accept mystery and then seek to clear it up. Within every new avenue that's cleaned up you get a fresh vista into the world beyond.'[4]

These were the outlines of ideas that were to be formalised thirty years later by the philosopher Karl Popper. Only two years after the war, Fry was still troubled by the mystery surrounding science, and he worried that it was holding back a developing confidence in science: 'We still have the method of science but we are losing for the time its faith.'[5] He was talking about himself and his friends, members of the establishment who were only slowly coming to terms with their different place in the new society.

It was not yet clear what the legacy of the Bloomsbury artists was to be. There were some very good books and paintings, even sculptures and fine pieces of furniture. As well as being the end of the Victorian age, many of the art pieces were expressions of anger at the war and its losses. Fry, Lytton Strachey and Virginia Woolf all dipped their toes into the scientific waters and then withdrew. They were confused about where science would take them, and unsure that they could afford to make a living as scientists. Was it partly as a result of this fear that they behaved in such a self-sufficient manner and did not look very far outside their own society?

The fear was not of science itself but of its attendant baggage, such as a meritocracy, which threatened their social position. Woolf, in particular, strongly disliked the new liberal middle class: she feared they would displace much of what she held dear with a monstrous mediocrity, a mass-produced materialism that challenged human decency. So throughout the 1920s she set up a set of written attacks on popular writers such as H. G. Wells, Arnold Bennett and John Galsworthy, and persuaded her like-minded friends, such as T. S. Eliot and D. H. Lawrence, to do the same. Woolf's own manifesto on this position was her twenty-four-page essay read to the Heretics Society in Cambridge.[6] More recently, a storm from very different contrasts blows up in Edward Albee's 1962 'Who's Afraid of Virginia Woolf?'

***

To everyone in Europe, and to many beyond, the war had emphasised the fragility of human life and showed that science could so easily influence its quality, one way and the other. For many, that meant support for society to control quality by a programme of eugenics. Already in Bloomsbury, the Eugenics Education Society was well established, and there was the political will to legislate, but science was moving on quickly and its social mission was spreading even more strongly in North America and Germany. But where could a eugenic programme of applied projects even begin?

During the 1920s, one of the few important breakthroughs in biology was an understanding of the pace of evolution. It was unusual for biological research to have an impact on a contemporary political issue. Questions of that sort niggled at the minds of politicians and legislators because unstable foundations for any new laws about eugenics did not augur well for easy enforcement. For example, there was disagreement about whether evolution occurred quickly by mutations or through slower, more continuous variation. The matter was briefly resolved by the leaders of a new specialism, population genetics: Ronald Fisher and J. B. S. Haldane. Their mathematics showed that small mutations were more effective than big ones in bringing about useful change within a population. All that was needed to explain that change were small alternative forms of the same gene. It followed from this that the supporters of Mendel were right in assuming that Darwinism needed a particulate mechanism for inheritance, that mutations took place within the cell and that selection happened outside each organism, in populations and their environment. But legislation on what might come from this was going to be difficult to draft, let alone implement.

But Fisher's work in this area did refine the way people thought about earlier theories. The cleric and economist Thomas Robert Malthus had argued a hundred years earlier that numbers of individuals within a species peaked before falling to extinction, but no one knew why or how. Now Fisher made a strong link between biology and physics in order to suggest that a large population contained more variation and so it had a larger chance of survival. The limit to its growth was usually provided by the amount of food available in a particular environment. But there were other causes of extinction outside Fisher's expertise. Haldane, for example, followed Darwin's view that evolutionary change could also happen by means other than natural selection. Haldane urged people to keep an open mind. For example, he had plenty of evidence that degeneration is a more common phenomenon than progress and is usually hard to spot because it also leads to extinction. There was also hybridisation and some large mutations that could make new species; from the fossil record, when we find one or two lines leading to extinction, dozens of others can lead into fresh directions.

Fisher's and Haldane's efforts were making some progress in explaining evolution even though its molecular mechanisms were not to be discovered until the 1960s. Nevertheless, very different attitudes to genetics experiments were developing after the war. These were set out by Ronald Fisher in 1930. His book *The Genetical Theory of Natural Selection* described how some genes are dominant and others recessive. It reported on many experiments of plant breeding in which statistical analysis of the characters from particular genes showed that they controlled characters such as petal colour and leaf shape. It was his first summary of this view of evolution and was to be followed through the next decade with more compositions by other authors with the same emphasis on the importance of mutations.

As a statistician Fisher argued that the pattern of social degradation in the years of economic depression arose from stress in the birth rate, not the death rate. He contended that, as a consequence, eugenic control needed to be made at birth. This was where his politics fixed to his biology; this was also where his reputation was buried. He frightened off supporters with propositions for extreme controls on births. Wanting to be his own master after the First World War, he rejected a suggestion from Pearson that they work together. He quickly found his own job at the Rothamsted Experimental Station near St Albans.

Fisher and Haldane came from similar backgrounds, were a similar age, attended similar schools, both went to Oxbridge, and both ended up in the late 1920s and early 1930s working with biometry at UCL. Despite these similarities, they argued incessantly, pursued different paths and held different ideas about politics and religion. The former was a right-wing Christian and the latter an atheist far to the left. Fisher saw God as a benign casino owner with a 'design by chance' policy, challenging humanity to work together by self-discipline to save the planet. Such a view made it easy to explain natural selection using probability theory and enabled him to apply self-discipline to fascist targets. This meant that Fisher was increasingly ignored through the 1930s and was only able to work on a few small projects at Rothamsted.

The last hundred pages of Fisher's *Genetical Theory of Natural Selection* was less about the detail of that title and more about selective breeding to 'improve human stock'. The chapter headings summarised the main topics: Man and Society, Inheritance of Human Fertility, Reproduction in Relation to Social Class, The Social Selection of Fertility.

Marie Stopes had also joined the Eugenics Education Society in 1912 and argued against one of the best known opponents of eugenics, Halliday Sutherland. He was a Marylebone physician and held open-air clinics at the Regent's Park bandstand especially to diagnose tuberculosis. Stopes sued Sutherland in 1923, and he won £100 damages. Six years later, he sued Stopes about an article in *Birth Control News*. That time, he lost.

Stopes wrote in the final chapter of her 1920 book *Radiant Motherhood* about her 'ardent dream' that science could offer different clinical techniques to reduce the size of families. One of her admirers up in Cambridge was William Bateson whose ideas about the role of single genes were becoming widely accepted. Particular illnesses and structural traits were being associated with particular genes. The work was being done on experimental animals and plants, but some hoped to experiment more adventurously. Bateson was becoming worried that his friend Erwin Baur, the German psychiatrist, was working in a society that was persuading him to try out some of the tests on humans.

***

As he grew into old age, Lankester's moods did not improve, but at least his many bad-tempered exchanges were still punctuated with an occasional brilliant insight. He was infuriated that the Kaiser Wilhelm was allowed to abdicate the Prussian throne and go to the Netherlands in exile. Lankester wrote to *The Times* that it was 'a perpetuation of the privilege accorded to one another by royal criminals, however great their responsibility for useless bloodshed'.[7] Things were made more difficult by Lankester's failing health, and, though he began to moderate his drinking and smoking, he found it hard to walk and became more reliant on his housekeeper in Chelsea, Miss Pearson. Miss Pearson's job was not easy. Lankester was a difficult man who lost his temper at the slightest disturbance. He lived in the large first-floor rooms and wrote at his big desk with a heavy paperweight shaped like a gorilla's foot, while on the mantelpiece behind was a picture of his old friend Anna Pavlova. If a book or scientific manuscript were moved from where he left it, all hell broke loose.

Some of Lankester's unpleasant behaviour was a way of hiding his anger about the war, a common reaction then in men of his age. One sure way of making him lose his temper was to cite new examples of spiritualism and the occult. When a new book *History of Spiritualism* was published in 1926, Lankester wrote a vicious criticism in the *Sunday Times*.

It didn't matter to him that the author was Arthur Conan Doyle, his old friend from Sussex who had been involved with the Piltdown fossils. To Lankester, the book was full of unscientific rubbish, misguiding the public into false ideas. Then there was another friend, Sir Oliver Lodge, a physicist who also practised 'psychic science'. Lankester would say to Lodge, 'You old charlatan, how are the spooks?'[8] Like fellow Ghost Club member, Arthur Conan Doyle, Lodge had lost a son in the war and turned to spiritualism soon after.

In the 1920s, Lankester wrote several important and brilliant summaries of the state of progress in biology. These showed that he was keeping up to date with the main developments: dating historical events by changes in tree-ring growth, photosynthesis, victory for the Creationists at the Scopes trial in Tennessee, Alexander Fleming's first recognition of penicillin. These outlines were followed by more complaints about the biological activities in universities in post-war society. In a letter to H. G. Wells, he wrote, 'The present biological activities in universities are reduced to rather feeble laboratory notebooks with curves and mathematical swagger about rates of movement, leading to nothing. No binding theory.'[9] As usual, he was right. In the 1920s, there was nothing holding the biological disciplines together, no overriding theory to test or goal to work for. Biology had become fragmented and specialised, and Lankester lamented this.

After the war, Wells was one of the people Lankester saw regularly. Lankester was too ill to attend the funeral of the biologist Jane Wells in 1927, so he grieved her early death alone. In contrast to the funerals of Charles Darwin and Karl Marx almost fifty years earlier, those who attended Jane Wells's funeral showed a broad mix of social class and cultures. So much had happened socially and politically, but the understanding of evolutionary biology was not much different. Shaw wore an orange handkerchief, Wells a blue overcoat. Virginia Woolf wrote in her diary, 'Poor Jane. It was desperate to see what a dowdy shabby imperfect lot we looked.'[10]

By the 1920s, there were several Bloomsbury artists and scientists working on projects that required them to think not just about abstract science but science in society and the science of society. Arthur Tansley, for instance, was proving in his work on the Norfolk coastal dunes and marshes that large systems in nature relied on interaction and cooperation. He was also thought to be helping the psychoanalysts Ernest Jones and Sigmund Freud to establish scientific principles to their methods. Marie Stopes was applying birth-control methods to help social hardship. Aldous Huxley talked about the political difficulties of looking after the environment. J. B. S. Haldane was advising the Russian Academy of Sciences on how to grow more crops.

Meanwhile, Lankester's health deteriorated. One of his last recorded comments was that 'I forget even the most interesting things for want of hearing them spoken of.'[11] He said this to a visitor from the British Museum who talked with him about freshwater medusae for over an hour. He died on 15 May 1929 at the age of eighty-two. The funeral was held at Saint Martin's in the Fields where the congregation sang 'Abide with Me' and listened to Chopin's *Marche Funebre*.

An epitaph to Ray Lankester's life and work came in the form of his friendship with H. G. and Jane Wells. It had been the source of great satisfaction to Lankester to encourage the career of the Wells's zoologist son, George Philip (known as Gip). Lankester had met Gip when he visited the Wells's home in Great Dunmow. They shared an interest in the local natural history, the invertebrates in particular. Gip succeeded in his more formal studies of biology where his father had failed. Not only had both of his parents been professional teachers of biology, they both doted on their first son and gave his education high priority. With a first-class degree in zoology from Cambridge in 1924, Gip went on to conduct research on lugworms at the Marine Biology Research Station in Plymouth, the station that Lankester had founded in the 1880s.

Gip proved himself a promising young scientist and gained an assistantship in zoology at UCL in 1928. There again he followed Lankester with a charismatic teaching style and personable role as a laboratory demonstrator. His particular interest was to link the results from physiology experiments in the laboratories of Gower Street with animal life in the environment around Plymouth. These kinds of comparisons had never been made before. They linked structure to function, behaviour to environment: a central essence of the evolutionary process. When his father persuaded him to join Julian Huxley in writing the *Science of Life*, it put a stop to his experiments. He understood that this was an unusual opportunity to form a working relationship with one of the leading biologists. A strong bond developed between the two men. *Science of Life* became a world standard text from 1931, when it was first published as a single volume, until the 1950s. It was free of jargon, so much so that it told of how green plants synthesise sugars and other energy from sunlight: without mentioning the word 'photosynthesis'. In the same vein, it described coal and oil as 'bottles of sunshine', one of the first public warnings that using fossil fuels is bad for the planet.

Gip Wells stayed on at UCL until the 1960s, enhancing his reputation as a skilled teacher and science populariser, and was elected to the

Royal Society in 1955. A few weeks before he died in 1985 he attended a Royal Society soirée, carrying a gold-mounted walking stick, inherited from his father. It was the same stick that had been given to H. G. Wells by Ray Lankester.

***

In 1927, two years after his appointment as professor of zoology at King's College, Julian Huxley began to contribute to *Science of Life*. Under pressure from Wells to increase his work for the project, Huxley did what few newly appointed professors ever do: he resigned. He was never to return to academic life. He preferred the independence of a freelance writer's life. He decided working in a group for less money was preferable to institutionalised academia. Wells strongly approved of his co-author's commitment to their encyclopaedia project, not least because it freed him to devote valuable time to other business.

Huxley now had the freedom to concentrate on another project, that of bringing together all the important new ideas about how evolution worked. There was a vast amount of work going on in such fields as genetics and biochemistry. New ways of dating geological processes enlivened palaeontology. Population genetics was just beginning. Numerous examples of evolution actually happening were crying out for comparison with one another and with changing environments. Julian Huxley was just the man to join these ideas together into a meaningful whole, and he started to add his own creative twists to how this might be done.

After a year of writing on the first project, and still being pushed by H. G. Wells to finish his contributions, Julian Huxley and his French wife Juliette decided to leave London for the long winter of 1928 and stay at the Swiss skiing resort of Les Diablerets. There they rented a chalet with Aldous Huxley and his wife Maria. Aldous also had a deadline to finish *Point Counter Point*. To add to the literary atmosphere at their remote Alpine encampment, the Huxleys' friends D. H. and Frieda Lawrence had a chalet nearby, where he was busy composing the final draft of *Lady Chatterley's Lover*.

The completion of these three works, by six good friends together at the same retreat, could not have been more appropriate and timely. In the mountains they were each trying to perceive the physical influences that comprised life, the basic biochemistry and physiology over which they had no control. They were trying to describe the interaction of these things and to bring them all together. Emotionally and intellectually exhausted by this work during the day, each evening they took turns to read aloud to one another and finished the whole of *The Pickwick Papers*. Julian saw it as 'a happy time, the white landscape soothing and protective, and much work was done'.[12]

According to contemporary accounts, the Huxleys enjoyed talking about their different evolutionary and physiological ideas, how species can thrive in extreme environments and how humankind developed genetically. These discussions infuriated Lawrence who still insisted that the more power that was exercised by 'the dark loins of man' the greater would be the freedom for our instincts and our intuitions. Julian Huxley saw evolution as a natural process, an opportunity for the fittest. Lawrence saw it as a challenge for the individual, for whom such a utopia was a singular and physical climax. For him, it was full of passion and desire, like the moon, 'a globe of dynamic substance, like radium or phosphorus, coagulated upon a vivid pole of energy'.[13] Aldous Huxley was infuriated by Lawrence's Bergson-like thinking, which went against all the latest evidence for natural selection. Aware of Lawrence's obstinacy, the Huxleys chose not to argue with him, for fear of disturbing the peace.

The group walked and skied in the mountainous forests, overcome by the silence and timelessness of their alpine retreat. 'She drips herself with water' wrote Lawrence in his poem *Gloire de Dijon*. Here was life in an extreme environment, and here were six people sharing intimate feelings, expressing themselves at the cutting edge of their varied expertise and interests, the Huxleys with their humanism, Lawrence with his scepticism about science and his trust in nature. Lawrence summed up their arguments with a poem he called 'Relativity':

> I like relativity and quantum theories
> because I don't understand them
> and they make me feel as if space shifted about like a swan that
>   can't settle,
> refusing to sit still and be measured;
> and as if the atom were an impulsive thing
> always changing its mind.[14]

These winter dialogues foreshadowed many of the new ideas that were to trouble intellectuals over the next few years. The Huxley brothers had a broad vision of biology and its hierarchy of scale: organs, tissues, cells, chromosomes and molecules. They understood that somewhere within this array each organism had a particular sense of self, allowing different parts of the hierarchy to have special responses both inwards and outwards to other parts. They disagreed with Lawrence about what is uniquely human and what is evidence for universal biological instinct. This had been at the centre of their alpine conversations, and showed up in many ways in their work. Despite the differences, a strong friendship developed between Lawrence and Aldous Huxley.

Having seen so much social change and human conflict, many in Bloomsbury strived to understand the post-war physical and psychological debris. Aldous Huxley, Lawrence and Virginia Woolf turned their attention to those survivors of the war who were trying to work through their losses. One of the most angry of the books from the time was Aldous Huxley's *Brave New World*, published in 1932. It was well received by both literary and scientific critics. Pointing to the authoritarian regimes of Soviet Russia and Nazi Germany, the novel warned of a totalitarian future. Huxley's imagined future was frightening because it was so credible. He made sure that the science that underpinned this brave new world, with its biological mechanisms and eugenic controls, was possible as well as plausible.

In the same year, Julian Huxley was appointed secretary of the Zoological Society in The Regent's Park. He went on to speculate about human responsibility for the environment and how one-world humanism might help to protect the environment. Both Huxleys expected that science would provide the key to ending the economic depression of the early 1930s. They thought it was up to them and their friends to advise the politicians on how to solve the problems created by the crisis.

As a first step, Gip Wells and the Huxleys invited some of their friends to form a small dining club. They called this club the Tots and Quots, after *quot hominei, tot sententiea*: so many men, so many opinions. Tots and Quots met for the first time in 1931 at Pagani's restaurant in Great Portland Street. The group included Solly Zuckerman, a young research assistant from the zoo who was involved in investigating racial differences within monkeys, apes and humans; another zoologist, J. Z. Young; the geneticists Lionel Penrose, Lancelot Hogben, Joseph Needham and J. B. S. Haldane; the crystallographer, J. D. Bernal; and the economist, Hugh Gaitskell.

They met in a room on the second floor of the restaurant with its ruby velvet curtains and mantel drapes. On the wall were brown squares of paper with drawings and writing protected by glass: songs, praises to the chef and other remarks by happy patrons. From their meetings in that Bloomsbury restaurant emerged not the expected political manifesto but an important book about evolutionary mechanisms, Julian Huxley's *Evolution: The Modern Synthesis*, published in 1942. The preface acknowledged the members of that group for their part in bringing together so many concepts about evolution from different disciplines, leading to the synthesis itself. The book brought together elements from anatomy, genetics, physiology, ecology and palaeontology and discussed how all of these disciplines have a bearing on adaptation and evolution by natural selection. It is generally regarded as the most important piece of work in biology of the first half of the twentieth century.

More advances in evolutionary thought came in 1944 when the Nobel Laureate in Physics Erwin Schrodinger gave a series of talks asking 'What Is Life?' In these talks Schrodinger predicted some of the forthcoming prospects of molecular biology. A few years later, in Drury Lane – one of the streets that Virginia Woolf used to walk from Bloomsbury to Soho – Maurice Wilkins and Rosalind Franklin took the famous x-ray photograph of crystalline DNA. It led to a new battle between Bloomsbury and Cambridge scientists, and once again ended controversially with James Watson and Francis Crick claiming victory with their 1953 *Nature* paper. It described the double helix structure of the DNA molecule. It was a triumph for the meritocracy and for Darwinism.

Most Bloomsbury biologists and artists were not the professional specialists we know today. They worked for a senior professor, observing, interpreting and describing, with words, drawings and experiments, but rarely with measurements and calculations. After the First World War, they did become more experimental and quantitative, and they were recruited by a growing meritocracy to have a profession with a career. But it was a few decades before another major switch in the way scientists work: when they began to falsify theories rather than prove them to be correct.

These Bloomsbury pioneers dominated the intellectual life of London for sixty years, coming between the gentlemen scientists, with laboratories in their own homes, and the competitive system we know today. They were pluralists with more ambition for society than for themselves and with much less attention to detail than is given today. That was because they were opening the fundamentals of so much knowledge, shallow at first and not deepening until they had more confidence and more sophisticated methods. They were professional pioneers through an age when science and society were very different shapes than they are now.

Postscript

As my train to Leicester pulls out of St Pancras station, a startling new building has just risen from the wasteland behind the British Library, its roof like the carapace of a giant insect. This is the Crick Institute. The Crick Institute was established as a complex hothouse of global inter-disciplinary teams. As the website says, 'Its work will help understand why disease develops and find new ways to treat, diagnose and prevent illnesses such as cancer, heart disease and stroke, infections, and neu-rodegenerative diseases … by bringing together scientists from all disci-plines.' The hope is that 'it will not only help to improve people's lives but will also keep the UK at the forefront of innovation in medical research, attract high-value investment, and strengthen the economy'.[1]

My mind slips back to 1961 when I first arrived in Bloomsbury, near this spot, and I think about how I understood biology then. Replacing the familiar view of great gasholders, car parks and grimy tenement blocks, a new multi-billion meritocracy is growing up here right in the heart of London. I quickly realised when I arrived in Bloomsbury that something special was going on there, something unexpected and surprising, and full of hope. Is the same happening again with this new building? Right here in Bloomsbury, scientists will find the codes of newly recognised genes.

The biology I learnt in Bloomsbury was startling and fascinating. It was the new molecular biology that had its roots in physics and chem-istry. The geology being discovered then, about drifting continents and the ages of extinct and living species, was unexpected too, and just as empirical. These and other avenues of knowledge were made possible by x-ray crystallography, radiometric dating and other techniques perfected during the Second World War.

Francis Crick was born in 1916 near Northampton, where his father had a small boot and shoe factory. At the bottom of his uncle's garden was a wooden shed where Francis learnt to develop photographic plates and practise glass blowing. He went to grammar school in Northampton, won a scholarship to Mill Hill School in north London in 1930 when he was fourteen, and studied physics at UCL from 1934. Without Latin, he wasn't eligible to apply for a place at Oxford or Cambridge, but his experience in Bloomsbury was rich enough for him to avail himself of big opportunities. A German bomb destroyed his x-ray tube in 1941. He was in the process of looking at the structure of protein molecules for a PhD. After the war, he moved on to Cambridge. Although his discovery of the double helical structure of deoxyribose nucleic acid happened in the laboratories of Cambridge, he was essentially a Bloomsbury man.

The development of London life sciences was being driven by sudden advances in technology, new machines that could accurately work out a substance's chemical composition, magnify the contents of a cell many thousand times and say how many million years ago a piece of rock was formed. Microscopy, spectroscopy, crystallography and new experimental methods had quickly and unexpectedly changed the understanding of cell biology. Together, these technologies allowed insights of the connections highlighted in Julian Huxley's *Modern Synthesis*: how genes function, how cells work and how species migrate around the planet. By filling in the details of these three parts of evolutionary studies, Darwin's theory of adaptation by natural selection was eventually proved beyond reasonable doubt.

This final proof of Darwinism came through the 1960s and 1970s, when I was teaching life sciences in a small London college. The jigsaw of information from the big research laboratories was slowly beginning to fit together. The information showed us that there is just one evolutionary tree for the DNA of all species, one animated migration map for all species across the planet. That principle is now clear and leaves just the details for future generations of biologists to figure out. Even more impressively, that same tree and that map can be traced back through time to the beginning of life, and that life has a beauty comprising art and science. From Leicester, C. P. Snow challenged artists with the second law of thermodynamics: disorder increases in a closed physical system. Now in Bloomsbury, with the rise of information in an open biological system, we know that biological evolution also has a broad scope. The early twentieth-century Bloomsbury biologists and artists had important ideas about evolution within their single culture. It is unscrupulous and opportunistic, never missing a chance to adapt to reproduce more members of the species in the new conditions, giving exuberant purposeless living: no plan, no end, no winners. The play is beautiful and elegant and knows no end.

The acres of rejuvenated space from the old railway lands at King's Cross and St Pancras are now home to the new University of the Arts London, King's Place Concert Hall, offices for the *Guardian* and *Nature* magazine, as well as to the Francis Crick Institute and the British Library. They are a northern extension to the Bloomsbury square mile and take off from the cultural achievements made there through the past two centuries.

The fundamental mysteries of evolution's hereditary mechanisms are now well known, based on an understanding of how DNA codes for particular amino acids, and how the chemical processes of epigenetics sense the environment and turn cell processes on and off. Evolution by natural selection is the mainstay of species diversity, an opportunistic process of self-organised adaptation to environmental change. The questions that were asked by the Bloomsbury scientists a hundred years ago have been amply answered. Galton and Pearson would have been pleased to find that many of the genetically transmitted illnesses they sought to eradicate with eugenics can now be investigated with real insight of their molecular and genetical causes. That is a major task of the Francis Crick Institute.

However, another worry of the nineteenth- and twentieth-century biologists remains: that from the earlier theologian Thomas Robert Malthus, that the natural rate of increase in the human population would reach a level of unsustainability. Now it is the single most important problem facing the natural world and is connected with many others: climate change, biodiversity loss, racial interaction, diminishing natural resources and pollution. The extinctions that result will include that of our own species and is already making other irreversible changes. It is a situation not conceived by the utopians such as H. G. Wells and Aldous Huxley, let alone Darwin. They all knew of Malthus's warning and were deeply influenced by its challenges, but they all had hope that science would find a solution, as do many of our own contemporaries. We are still hoping.

Notes

Introduction

1   C. P. Snow, *Homecomings* (Harmondsworth: Penguin, 1962), p. 155.

1 Two funerals, 1882–3

1   T. Hunt, *The Frock-Coated Communist: The Revolutionary Life of Friedrich Engels* (London: Allen Lane, 2009), p. 288.
2   Hunt, *The Frock-Coated Communist*, p. 288.
3   J. Bentham, *An Introduction to the Principles of Morals and Legislation*, 2nd ed. (Oxford: Clarendon Press, 1823), p. 147.
4   V. Woolf, *The Years* (Harmondsworth: Penguin, 1968), p. 59.
5   N. Annan, *Leslie Stephen: The Godless Victorian* (Chicago, IL: Chicago University Press, 1984), p. 91.
6   C. Darwin, *The Descent of Man* (London: John Murray, 1879).
7   Annan, *Leslie Stephen*, p. 108; also T. and F. Hardy, *The Life of Thomas Hardy, 1840–1928* (Ware: Wordsworth Editions, 2008).
8   Annan, *Leslie Stephen*, p. 140.
9   Annan, *Leslie Stephen*, p. 111.
10 G. Eliot, *Middlemarch* (Oxford: Oxford University Press, 1871–2), p. 3; V. Woolf, *Times Literary Supplement*, 20 November 1919.
11 M. Holroyd, *Lytton Strachey: A Biography* (Harmondsworth: Penguin, 1971), p. 13.
12 Holroyd, *Lytton Strachey*, p. 32.
13 M. Brookes, *Extreme Measures: The Dark Visions and Bright Ideas of Francis Galton* (London: Bloomsbury, 2004).
14 N. W. Gillham, *A Life of Sir Francis Galton* (Oxford: Oxford University Press, 2001), p. 270.

2 Lankester takes over, 1884–90

1   J. Lester, *E. Ray Lankester and the Making of Modern British Biology*, ed. P. J. Bowler (Faringdon: British Society for the History of Science, 1995), p. 51.
2   Lester, *E. Ray Lankester*, p. 51.
3   C. Darwin to R. Lankester, 11 April 1872, in F. Darwin and A. C. Seward, eds., *More Letters of Charles Darwin*, vol. II (London: J. Murray, 1903), p. 336.
4   Olive Schreiner met Ray Lankester at a dinner party in 1881 (Lankester Family Papers).
5   J. Endersby, *A Guinea Pig's History of Biology* (London: Heinemann, 2007), p. 140.
6   R. J. Richards, *The Tragic Sense of Life* (Chicago, IL: Chicago University Press, 2008), Chapter 5, p. 113–70.
7   Appendix by Karl Marx to Ray Lankester's notes on degeneration, perhaps added for a lecture (Lankester Family Papers).

3 Eccentric campaigners, 1890s

1. M. Holroyd, *Bernard Shaw*, 3 vols. (Harmondsworth: Penguin, 1988), vol. I: 1856–98, p. 99, describing Shaw's social awkwardness at the Savile Club.
2. G. B. Shaw, *Back to Methuselah* (Oxford: Oxford University Press, 1921), p. 32.
3. Shaw, *Back to Methuselah*, p. 26.
4. Shaw, *Back to Methuselah*, p. 27.
5. Shaw, *Back to Methuselah*, p. 27.
6. Porter, *Karl Pearson: The Scientific Life in a Statistical Age* (Princeton, NJ: Princeton University Press, 2004), p. 1.
7. T. M. Porter, *Karl Pearson*, p. 1.
8. K. Pearson to Robert Parker, 24 October 1881, in Porter, *Karl Pearson*, p. 135.
9. J. Browne, *Charles Darwin: The Power of Place, Volume 2 of a Biography* (London: Pimlico, 2003), vol. II, p. 355.
10. S. Butler, 'The Deadlock in Darwinism', in R. A. Streatfield, ed., *Essays in Life, Science and Art* (New York, NY: Kennikat Press, 1970), pp. 234–340, at p. 243.
11. Butler, 'The Deadlock in Darwinism', p. 25.
12. C. G. Stillman, *Samuel Butler: A Mid-Victorian Modern* (London: Martin Secker, 1932), p. 205.
13. Stillman, *Samuel Butler*, p. 205.
14. Lester, *E. Ray Lankester*, p. 115.
15. A. Desmond, *Huxley: The Devil's Disciple*, 2 vols. (London: Michael Joseph, 1997), vol. II, p. 171.
16. Desmond, *Huxley*, vol. II, p. 172.
17. Desmond, *Huxley*, vol. II, p. 172.
18. W. Booth, *In Darkest England and the Way Out* (London: Salvation Army, 1890).
19. Lester, *E. Ray Lankester*, p. 42.

4 Insiders and outsiders, 1890s

1. V. Woolf, *Roger Fry: A Biography* (London: The Hogarth Press, 1940).
2. V. Brome, *H. G. Wells: A Biography* (London: Longmans, 1951), p. 15.
3. Brome, *H. G. Wells*, p. 33. V. Woolf, *To the Lighthouse* (London: The Hogarth Press, 1927), p. 98.
4. The origin of this is uncertain. One clue is that the poet Henry Newbolt and the philosopher John McTaggart were contemporaries at Clifton College.
5. A. Whitehead used these words when he introduced Bertrand Russell to an audience for the William James Lecture at Harvard in 1940.
6. P. Levy, *Moore: G. E. Moore and the Cambridge Apostles* (Oxford: Oxford University Press, 1981), p. 170.
7. Levy, *Moore*, p. 172.
8. P. Levy, *G. E. Moore and the Cambridge Apostles* (Oxford: Oxford University Press, 1979).
9. T. Carlyle, 'Signs of the Times', *Edinburgh Review*, June 1829.
10. Ray Lankester to H. G. Wells, 12 July 1900. Wells Papers, University of Illinois.
11. H. G. Wells to Ray Lankester, 12 or 14 July 1900, Lankester Family Papers.
12. Woolf, *Roger Fry*, p. 75.
13. Woolf, *Roger Fry*, p. 65.
14. Woolf, *Roger Fry*, p. 65.

5 A new breed of professional, 1890–1904

1. Ray Lankester, 4 October 1890. Lankester Family Papers.
2. Lester, *E. Ray Lankester*, p. 176.

3 V. Woolf, *To the Lighthouse* (London: The Hogarth Press, 1927), p. 98.
4 Quoted in M. C. Boulter, ed., 'Palaeobotanical Extracts from the *Sportophyte*', *International Organisation of Palaeobotany*, Circular 7 (1981).
5 C. Darwin, *On the Origin of Species by Means of Natural Selection* (Harmondsworth: Penguin, 1985), p. 458.

6 A new breed of biologist, 1900–10

1 A. E. Shipley, 'Walter Frank Raphael Weldon', *Proceedings of the Royal Society*, B 80 (1908), 25–41.
2 R. Weldon, 'Some Remarks on Variation in Plants and Animals', *Proceedings of the Royal Society* 57 (1895): 361–79, at p. 361.
3 W. Bateson, *Materials for the Study of Variation* (London: Macmillan, 1894), p. 320.
4 Galton Archive, UCL, List no. 198. William Bateson letters to Francis Galton.
5 Bateson, *Materials for the Study of Variation*, p. 320.
6 R. Skidelsky, *John Maynard Keynes, 1883–1946: Economist, Philosopher, Statesman* (London: Pan Macmillan, 2003), p. 135.
7 Skidelsky, *John Maynard Keynes*, p. 273.
8 G. B. Shaw to K. Pearson, 20 June 1893. Pearson Papers 627.
9 Porter, *Karl Pearson*, p. 1.
10 H. G. Wells, *Ann Veronica* (London: Fisher Unwin, 1909).
11 P. N. Furbank, *E. M. Forster: A Life* (London: Secker & Warburg, 1977), p. 162.
12 F. Spalding, *Gwen Raverat* (London: The Harvill Press, 2001), p. 173.
13 H. G. Wells, *The Time Machine* (London: Heinemann, 1895).

7 The rise of eugenics, 1901–14

1 Ray Lankester, Notes on degeneration, Lankester Family Papers.
2 Gillham, *A Life of Sir Francis Galton*, p. 331.
3 Gillham, *A Life of Sir Francis Galton*, p. 331.
4 Porter, *Karl Pearson*, p. 285.
5 Porter, *Karl Pearson*, p. 285.
6 F. Galton, *Essays in Eugenics* (London: Eugenics Education Society, 1909), pp. 1–34.
7 H. G. Wells, *Sociological Papers* (London: Macmillan, 1905), pp. 58–660; H. G. Wells, *A Modern Utopia* (London: Chapman & Hall, 1905).
8 F. Galton, *Noteworthy Families* (London: Macmillan, 1909), p. xli.
9 K. Pearson, *The Life, Letters and Labours of Francis Galton* (Cambridge: Cambridge University Press, 1930), vol. III, p. 658.
10 C. Brinton, *F. W. Nietzsche* (New York, NY: Harper & Row, 1965), p. 120.
11 Gillham, *A Life of Sir Francis Galton*, p. 273.
12 K. Pearson, 'The Cavendish Lecture', *West London Medical Journal* 17 (1912): 165–93.
13 Pearson, 'The Cavendish Lecture.'
14 Gillham, *A Life of Sir Francis Galton*, p. 346.
15 A. Balfour, 'First International Eugenics Congress', *British Medical Journal*, 3 August 1912, p. 253.
16 Balfour, 'First International Eugenics Congress', p. 253.
17 Balfour, 'First International Eugenics Congress', p. 255.
18 Gillham, *A Life of Sir Francis Galton*, p. 348.
19 Quoted in D. J. Childs, *Modernism and Eugenics* (Cambridge: Cambridge University Press, 2001), p. 27.
20 M. Kohn, *A Reason for Everything: Natural Selection and the English Imagination* (London: Faber & Faber, 2004), p. 91.
21 R. A. Fisher, 'Has Mendel's Work Been Rediscovered?' *Annals of Science* 1 (1936): 115–37.

8 Old habits die hard, 1901–14

1   Lester, *E. Ray Lankester*, p. 142.
2   Lester, *E. Ray Lankester*, p. 142.
3   Dinner at The Stage Society, 18 May 1913. Lester, *E. Ray Lankester*, p. 203.
4   Q. Bell, *Virginia Woolf: A Biography*, 2 vols. (London: The Hogarth Press, 1972), vol. I, p. 124.
5   M. Holroyd, *Lytton Strachey: A Biography* (Harmondsworth: Penguin, 1971).
6   C. Alt, *Virginia Woolf and the Study of Nature* (Cambridge: Cambridge University Press, 2010), p. 140.
7   V. Woolf, *Mr. Bennett and Mrs. Brown* (London: The Hogarth Press, 1924), pp. 4–5.
8   F. Spalding, *Vanessa Bell: Portrait of the Bloomsbury Artist* (London: Weidenfeld & Nicolson, 1983), p. 128.
9   P. Mini, *John Maynard Keynes: A Study in the Psychology of Original Work* (Basingstoke: Macmillan, 1994), p. 25; S. C. Dow and J. Hillard, *Keynes, Uncertainty and the Global Economy* (Cheltenham: Elgar, 2002), p. 41.
10 Quoted in V. Woolf, *Virginia Woolf: Selected Essays*, ed. D. Bradshaw (Oxford: Oxford University Press, 2008), p. 35.
11 Woolf, *Roger Fry*, p. 154.
12 Woolf, *Roger Fry*, p. 155.
13 L. Edel and G. N. Ray, *Henry James and H. G. Wells* (London: Hart-Davis, 1958).
14 V. Woolf, *A Writer's Diary*, ed. L. Woolf (Orlando, FL: Harcourt, 1953), 16 August 1922.
15 *Daily Mirror*, 30 March 1914.

9 Time passes, 1914–18

1   H. G. Wells, *The World Set Free: A Story of Mankind* (London: Macmillan, 1914), p. 1.
2   Lester, *E. Ray Lankester*, p. 205.
3   Lester, *E. Ray Lankester*, p. 206.
4   H. Gaudier-Brezska, *Vortex Gaudier-Brezska* (Santa Barbara, CA: Black Sparrow Press, 1981), July 1915, *Blast* 2, p. 34.
5   Kohn, *Reason for Everything*, p. 150.
6   Fisher, 'Has Mendel's Work Been Rediscovered?'
7   H. Godwin, *Biographical Memoirs of Fellows of the Royal Society* 3 (1957): 227–46.
8   Godwin, *Biographical Memoirs*, p. 230.
9   J. Huxley, *Memories, Vol. I* (Harmondsworth: Penguin, 1972), p. 228.
10 J. Huxley, *Leaves of the Tulip Tree* (London: John Murray, 1986), p. 101.
11 Huxley, *Leaves of the Tulip Tree*, p. 101.
12 G. Woodcock, *Dawn and the Darkest Hour: A Study of A. Huxley* (Montreal: Black Rose Books, 2007), p. 124.
13 D. H. Lawrence, *Sons and Lovers* (London: Duckworth, 1913), p. 137.
14 H. G. Wells, *Boon* (London: Faber, 2008), p. 72.

10 The one culture, 1920s and 1930s

1   Shaw, *Back to Methuselah*, p. 36.
2   S. P. Rosenbaum, *The Bloomsbury Group Memoir Club* (Basingstoke: Palgrave Macmillan, 2014), p. 112.
3   Woolf, *Roger Fry*, p. 270.
4   Woolf, *Roger Fry*, p. 271.
5   Woolf, *Roger Fry*, p. 271.
6   Woolf, *Mr Bennett and Mrs Brown*.
7   Ray Lankester, 13 November and 11 December 1918, to *The Times*.

8   Lester, *E. Ray Lankester*, p. 213.

9   Lester, *E. Ray Lankester*, p. 211.

10 V. Woolf, *The Diary of Virginia Woolf, 1925–1930*, ed. A. O. Bell (London: The Hogarth Press, 1980), p. 164.

11 Lester, *E. Ray Lankester*, p. 215.

12 J. Huxley, 'H. G. Wells', in J. R. Hammond, ed., *H. G. Wells: Interviews and Recollections* (London: Macmillan, 1980), p. 80.

13 D. H. Lawrence, *Fantasia of the Unconscious* (New York, NY: Dover Edition, 2005).

14 D. H. Lawrence, *Pansies: Poems* (London: Secker, 1929).

Postscript

1   www.kcalondon.com/our-work/the-story-of-the-crick/ (accessed 11 May 2017).

Bibliography

Alt, C. *Virginia Woolf and the Study of Nature*. Cambridge: Cambridge University Press, 2010.

Anderson, G. *Hang Your Halo in the Hall: The Savile Club from 1868*. London: The Savile Club, 1993.

Anker, P. *Imperial Ecology: Environmental Order in the British Empire, 1895–1945*. Cambridge, MA: Harvard University Press, 2001.

Annan, N. *Leslie Stephen: The Godless Victorian*. Chicago, IL: Chicago University Press, 1984.

Antliff, M., and V. Greene, eds. *The Vorticists: Manifesto for a Modern World*. London: Tate Gallery, 2010.

Ashton, R. *Victorian Bloomsbury*. New Haven, CT: Yale University Press, 2012.

Balfour, A. 'First International Eugenics Congress', *British Medical Journal*, 3 August 1912.

Bashford, A., and P. Levine, eds. *The Oxford Handbook of the History of Eugenics*. Oxford: Oxford University Press, 2010.

Bateson, W. *Materials for the Study of Variation*. London: Macmillan, 1894.

Bedford, S. *Aldous Huxley: A Biography*. London: Chatto & Windus, 1974.

Bell, Q. *Virginia Woolf: A Biography*, 2 vols. London: The Hogarth Press, 1972.

Bentham, J. *An Introduction to the Principles of Morals and Legislation*, 2nd ed. Oxford: Clarendon Press, 1823.

Bergson, H. *Creative Evolution: Authorized Translation by A. Mitchell*. London: Macmillan, 1920.

Bernstein, S. D. *Roomscape: Women Writers in the British Museum from George Eliot to Virginia Woolf*. Edinburgh: Edinburgh University Press, 2013.

Booth, W. *In Darkest England and the Way Out*. London: Milton, 1890.

Boulter, M. C., ed. 'Palaeobotanical Extracts from the *Sportophyte*', *International Organisation of Palaeobotany*, Circular 7, 1981.

Bowler, P. J. *The Eclipse of Darwinism: Anti-Darwinian Evolution Theories in the Decades around 1900*. Baltimore, MD: John Hopkins University Press, 1983.

Bowler, P. J. *Darwinism Deleted: Imagining a World without Darwin*. Chicago, IL: Chicago University Press, 2013.

Brinton, C. F. W. *Nietzsche*. New York: Harper & Row, 1965.

Brome, V. *H. G. Wells: A Biography*. London: Longmans Green, 1951.

Brookes, M. *Extreme Measures: The Dark Visions and Bright Ideas of Francis Galton*. London: Bloomsbury, 2004.

Brown, C., ed. *Lucien Pissaro in England: The Eragny Press, 1895–1914*. Oxford: Ashmolean Museum, 2011.

Browne, J. *Charles Darwin: The Power of Place, Volume 2 of a Biography*. London: Pimlico, 2003.

Butler, S. 'The Deadlock in Darwinism', in R. A. Streatfield, ed., *Essays in Life, Science and Art*. New York, NY: Kennikat Press, 1970, pp. 234–340.

Caine, B. *Bombay to Bloomsbury*. Oxford: Oxford University Press, 2005.

Calder, J. *The Nine Lives of Naomi Mitchison*. London: Virago, 1997.

Carlyle, T. 'Signs of the Times', *Edinburgh Review*, June 1829.

Carnforth, M. *Dialectical Materialism: An Introduction*. London: Lawrence & Wishart, 1952.

Casti, J. L. *The Cambridge Quintet: A Work of Scientific Speculation*. Reading, MA: Perseus Books, 1998.

Childs, D. J. *Modernism and Eugenics: Woolf, Eliot, Yeats, and the Culture of Degeneration*. Cambridge: Cambridge University Press, 2001.

Childs, P., ed. *Modernism: The New Critical Idiom*. London: Routledge, 2000.

Collini, S. *C. P. Snow: The Two Cultures*. Cambridge: Cambridge University Press, 1998.

Corning, P. A. *Holistic Darwinism: Synergy, Cybernetics, and the Bioeconomics of Evolution*. Chicago, IL: Chicago University Press, 2005.

Crook, D. P. *Benjamin Kidd: Portrait of a Social Darwinist*. Cambridge: Cambridge University Press, 1984.

Darwin, C. *On the Origin of Species by Means of Natural Selection*. Harmondsworth: Penguin, 1985.

Darwin, C. *The Descent of Man*. London: John Murray, 1879.

Darwin, F. and A. C. Seward, eds., *More Letters of Charles Darwin*, vol. II. London: J. Murray, 1903.

Desmond, A. *Huxley: The Devil's Disciple*. 2 vols. London: Michael Joseph, 1997.

Dewey, J. *The Influence of Darwin on Philosophy*. New York, NY: Prometheus Books, 1997.

Dow, S. C. and J. Hillard, *Keynes, Uncertainty and the Global Economy*. Cheltenham: Elgar, 2002.

Drake, B., and M. Cole. *Our Partnership: Beatrice Webb*. London: Longmans, 1948.

Easterlin, N. *A Biocultural Approach to Literary Theory and Interpretation*. Baltimore, MD: John Hopkins University Press, 2012.

Edel L. and G. N. Ray, *Henry James and H. G. Wells*. London: Hart-Davis, 1958.

Edwards, P., ed. *Blast: Vorticism, 1914–1918*. Aldershot: Ashgate Publishing, 2000.

Eliot, G. *Middlemarch*. Oxford: Oxford University Press, 1871–2.

Endersby, J. *A Guinea-Pig's History of Biology: The Plants and Animals Who Taught Us the Facts of Life*. London: Heinemann, 2007.

Endersby, J. *Imperial Nature: Joseph Hooker and the Victorian Practices of Science*. Chicago, IL: Chicago University Press, 2008.

Fisher, R. A. *The Genetical Theory of Natural Selection*. Oxford: Oxford University Press, 1930.

Fisher, R. A. 'Has Mendel's Work Been Rediscovered?' *Annals of Science* 1 (1936): 115–37.

Fry, R. *Transformations: Critical and Speculative Essays on Art*. London: Chatto & Windus, 1926.

Fry, R. *Vision and Design*. London: Chatto & Windus, 1920.

Furbank, P. N. *E. M. Forster: A Life*. 2 vols. London: Secker & Warburg, 1977.

Galton, F. *Essays in Eugenics*. London: Eugenics Education Society, 1909.

Galton, F. *Noteworthy Families*. London: Macmillan, 1909.

Gaudier-Brezska, H. *Vortex Gaudier-Brezska*. Santa Barbara, CA: Black Sparrow Press, 1981.

Gillham, N. W. *A Life of Sir Francis Galton*. Oxford: Oxford University Press, 2001.

Glendenning, V. *Rebecca West: A Life*. London: Weidenfeld & Nicolson, 1987.

Godwin, H. *Biographical Memoirs of Fellows of the Royal Society* 3 (1957): 227–46.

Gould, S. J. *The Structure of Evolutionary Theory*. Cambridge, MA: Belknap Press of Harvard University Press, 2002.

Hale, P. J.. *Political Descent: Malthus, Mutualism and the Politics of Evolution in Victorian England*. Chicago, IL: Chicago University Press, 2014.

Hall, R. *Marie Stopes: A Biography*. London: André Deutsch, 1977.

Hardy, T. and F. Hardy, *The Life of Thomas Hardy, 1840–1928*. Ware: Wordsworth Editions, 2008.

Harris, A. *Romantic Moderns: English Writers, Artists and the Imagination from Virginia Woolf to John Piper*. London: Thames & Hudson, 2010.

Henig, R. M. *The Monk in the Garden: The Lost and Found Genius of Gregor Mendel, the Father of Genetics*. Boston, MA: Mariner Books, 2001.

Holmes, R. *Eleanor Marx: A Life*. London: Bloomsbury, 2014.

Holroyd, M. *Bernard Shaw*. 3 vols. Harmondsworth: Penguin, 1988.

Holroyd, M. *Lytton Strachey: A Biography*. Harmondsworth: Penguin, 1971.

Humphreys, R. *Wyndham Lewis*. London: Tate Publishing, 2004.

Hunt, T. *The Frock-Coated Communist: The Revolutionary Life of Friedrich Engels*. London: Allen Lane, 2009.

Huxley, J. *Evolution: The Modern Synthesis*. Oxford: Oxford University Press, 1942.

Huxley, J. *Evolutionary Humanism*. New York: Harper & Row, 1964.

Huxley, J. 'H. G. Wells', in J. R. Hammond, ed., *H. G. Wells: Interviews and Recollections*. London: Macmillan, 1980.

Huxley, J. *Leaves of the Tulip Tree*. London: John Murray, 1986.

Huxley, J. *Memories, Vol. I*. Harmondsworth: Penguin, 1972.

Huxley, J. ed. *Aldous Huxley, 1894–1963*. London: Chatto & Windus, 1965.

Huxley, J. ed. *The New Systematics*. Oxford: Oxford University Press, 1940.

Kapp, Y. *Eleanor Marx*. Vol. 1: *Family Life, 1855–1883*. London: Lawrence & Wishart, 1972.

Kapp, Y. *Eleanor Marx*. Vol. 2: *The Crowded Years, 1884–1898*. London: Lawrence & Wishart, 1976.

Kerr, J. *A Most Dangerous Method: The Story of Jung, Freud, and Sabina Spielrein.* London: Sinclair Stevenson, 1994.

Knights, S. *Bloomsbury's Outsider: A Life of David Garnett.* London: Bloomsbury Reader, 2015.

Kohn, M. *A Reason for Everything: Natural Selection and the English Imagination.* London: Faber & Faber, 2004.

Lankester, E. R. *Degeneration: A Chapter in Darwinism.* London: Macmillan, 1880.

Lankester, E. R. *Science from an Easy Chair.* London: Methuen, 1910.

Lawrence, D. H. *Fantasia of the Unconscious.* New York, NY: Dover Edition, 2005.

Lawrence, D. H. *Pansies: Poems.* London: Secker, 1929.

Lawrence, D. H. *Sons and Lovers.* London: Duckworth, 1913.

Lee, H. *Virginia Woolf.* London: Chatto & Windus, 1996.

Lester, J. E. *Ray Lankester and the Making of Modern British Biology*, ed. P. J. Bowler. Faringdon: British Society for the History of Science, 1995.

Levenson, M. *Modernism and the Fate of Individuality.* Cambridge: Cambridge University Press, 1991.

Levine, G. *Darwin and the Novelists: Patterns of Science in Victorian England.* Chicago, IL: Chicago University Press, 1988.

Levy, P. *Moore: G. E. Moore and the Cambridge Apostles.* Oxford: Oxford University Press, 1981.

Lewis, H. *Olive Schreiner: The Other Side of the Moon.* Cape Town: Ihihihili Press, 2010.

Lodge, D. *A Man of Parts: A Novel.* London: Harvill Secker, 2011.

McGarr, P. and S. Rose, eds. *The Richness of Life: The Essential Stephen Jay Gould.* London: Vintage Books, 2007.

MacGibbon, J. *There's the Lighthouse: A Biography of Adrian Stephen.* London: James & James, 1997.

Maddox, B. *Rosalind Franklin: The Dark Lady of DNA.* London: Harper Collins, 2003.

Matthews, S., ed. *Modernism: A Source Book.* Basingstoke: Palgrave Macmillan, 2008.

Mini, P. *John Maynard Keynes: A Study in the Psychology of Original Work.* Basingstoke: Macmillan, 1994.

Mitchison, N. *Change Here: Girlhood and Marriage.* London: The Bodley Head, 1975.

Monk, R. *Bertrand Russell: The Ghost of Madness, 1921–1970.* London: Vintage, 2001.

Ortolano, G. *The Two Cultures Controversy: Science, Literature and Cultural Politics in Post-War Britain.* Cambridge: Cambridge University Press, 2009.

Overy, R. *The Morbid Age: Britain between the Wars.* London: Allen Lane, 2009.

Pearson, K. *The Life, Letters and Labours of Francis Galton.* Cambridge: Cambridge University Press, 1930.

Pearson, K. 'The Cavendish Lecture', *West London Medical Journal* 17 (1912): 165–93.

Peel, R. A., ed. *Essays in the History of Eugenics: Proceedings of a Conference Organized by The Galton Institute, London, 1997*. London: The Galton Institute, 1998.

Poole, R. *The Unknown Virginia Woolf*. 3rd ed. Atlantic Highlands, NJ: Humanities Press, 1990.

Porter, T. M. *Karl Pearson: The Scientific Life in a Statistical Age*. Princeton, NJ: Princeton University Press, 2004.

Pryor, W., ed. *Virginia Woolf and the Raverats: A Different Sort of Friendship*. Bath: Clear Books, 2003.

Ramsden, E. 'Eugenics from the New Deal to the Great Society: Genetics, Demography and Population Quality.' *Studies in History and Philosophy of Biological and Biomedical Sciences* 39 (2008): 391–406.

Raverat, G. *Period Piece: The Cambridge Childhood of Darwin's Granddaughter*. London: Faber & Faber, 1952.

Richards, R. J. *The Tragic Sense of Life*. Chicago, IL: Chicago University Press, 2008.

Roiphe, K. *Uncommon Arrangements: Seven Portraits of Married Life in London Literary Circles, 1910–1931*. London: Virago Press, 2007.

Rosenbaum, S. P. *The Bloomsbury Group Memoir Club*. Basingstoke: Palgrave Macmillan, 2014.

Russell, B. *My Philosophical Development*. London: Routledge, 1959.

Russell, M. *Piltdown Man: The Secret Life of Charles Dawson and the World's Greatest Archaeological Hoax*. Stroud: Tempus Publishing, 2003.

Sackville-West, V. *In Your Garden*. London: Michael Joseph, 1951.

Schmidt, M. *The Novel: A Biography*. Cambridge, MA: Belknap Press of Harvard University Press, 2014.

Seymour, M. *Ottoline Morrell: A Life on the Grand Scale*. Sevenoaks: Hodder & Stoughton, 1992.

Shaffer, E. *Erewhons of the Eye: Samuel Butler as Painter, Photographer and Art Critic*. London: Reaktion Books, 1988.

Shaw, B. *Back to Methuselah*. Oxford: Oxford University Press, 1921.

Sherborne, M. *H. G. Wells: Another Kind of Life*. London: Peter Owen, 2010.

Shipley, A. E., 'Walter Frank Raphael Weldon', *Proceedings of the Royal Society*, B 80 (1908), 25–41.

Shone, R. *Bloomsbury Portraits: Vanessa Bell, Duncan Grant and Their Circle*. London: Phaidon Press, 1976.

Skidelsky, R. *John Maynard Keynes, 1883–1946: Economist, Philosopher, Statesman*. London: Pan Macmillan, 2003.

Smith, D. C. *The Correspondence of H. G. Wells*. 4 vols. London: Pickering & Chatto, 1998.

Snow, C. P. *Homecomings*. Harmondsworth: Penguin, 1962.

Spalding, F. *Gwen Raverat: Friends, Family and Affectations*. London: The Harvill Press, 2001.

Spalding, F. *Vanessa Bell: Portrait of the Bloomsbury Artist*. London: Weidenfeld & Nicolson, 1983.

Spalding, F. *Virginia Woolf: Art, Life and Vision*. London: National Portrait Gallery, 2014.

Stillman, C. G. *Samuel Butler: A Mid-Victorian Modern*. London: Martin Secker, 1932.

Stott, R. *Darwin's Ghosts: The Secret History of Evolution*. London: Bloomsbury, 2012.

Sturgis, M. *Walter Sickert: A Life*. London: Harper Collins, 2005.

Sulloway, F. J. *Freud: Biologist of the Mind*. London: Fontana, 1979.

Thale, J. *C. P. Snow*. Edinburgh: Oliver & Boyd, 1964.

Thwaite, A. *Edmund Gosse: A Literary Landscape*. London: Secker & Warburg, 1984.

Weiner, J. S. *The Piltdown Forgery: The Classic Account of the Most Famous and Successful Hoax in Science*. Oxford: Oxford University Press, 1955.

Weldon, R. 'Some Remarks on Variation in Plants and Animals', *Proceedings of the Royal Society* 57 (1895): 361–79.

Wells, H. G. *Ann Veronica*. London: Fisher Unwin, 1909.

Wells, H. G. *Boon*. London: Faber, 2008.

Wells, H. G. *Experiment in Autobiography*. 2 vols. London: Victor Gollancz, 1934.

Wells, H. G. *A Modern Utopia*. London: Chapman & Hall, 1905.

Wells, H. G. *New Worlds for Old*. London: Archibald Constable, 1908.

Wells, H. G. *The Outline of History: Being a Plain History of Life and Mankind*. New York, NY: Garden City Books, 1920.

Wells, H. G. *Sociological Papers*. London: Macmillan, 1905.

Wells, H. G. *The Time Machine*. London: Heinemann, 1895.

Wells, H. G. *The World Set Free: A Story of Mankind*. London: Macmillan, 1914.

Wells, H. G. ed. *H. G. Wells in Love: Postscript to An Experiment in Autobiography*. London: Faber & Faber, 1984.

Wells, H. G. ed. *Socialism and the Great State: Essays in Construction*. New York, NY: Harper & Brothers, 1912.

White, P. *Thomas Huxley: Making the 'Man of Science'*. Cambridge: Cambridge University Press, 2003.

Woodcock, G. *Dawn and the Darkest Hour: A Study of A. Huxley*. Montreal: Black Rose Books, 2007.

Woolf, V. *The Death of the Moth, and Other Essays*. London: Hogarth Press, 1942.

Woolf, V. *The Diary of Virginia Woolf, 1925–1930*, ed. A. O. Bell. London: The Hogarth Press, 1980.

Woolf, V. *Mr. Bennett and Mrs. Brown*. London: The Hogarth Press, 1924.

Woolf, V. *Roger Fry: A Biography*. London: Hogarth Press, 1940.

Woolf, V. *Virginia Woolf: Selected Essays*, ed. D. Bradshaw. Oxford: Oxford University Press, 2008.

Woolf, V. *The Waves*. London: Hogarth Press, 1931.

Woolf, V. *A Writer's Diary*, ed. L. Woolf. Orlando, FL: Harcourt, 1953.

Woolf, V. *The Years*. Harmondsworth: Penguin, 1968.

Worchen, J. *D. H. Lawrence: The Life of an Outsider*. London: Allen Lane, 2005.

Index

CPSIA information can be obtained
at www.ICGtesting.com
Printed in the USA
BVHW01*0940090318
509887BV00013B/248/P